Introduction
to
Genetic Engineering

William H. Sofer
Waksman Institute
Rutgers - The State University of New Jersey
Piscataway, New Jersey

Butterworth - Heinemann
Boston London Singapore Sydney Toronto Wellington

Library of Congress Cataloging-in-Publication Data

Sofer, William.
    Introduction to genetic engineering / William Sofer.
    p.   cm.
    Includes bibliographical references and index
    ISBN 0-7506-9114-X
    1. Genetic Engineering. I. Title.
QH442.S65 1991
575.1' 0724 – – dc20

90-26127
CIP

British Library Cataloguing in Publication Data

Sofer, William
    Introduction to genetic engineering / William Sofer.
    I. Genetic engineering
    I. Title
    660.65

    ISBN 0-7506-9114-X

Butterworth-Heinemann
80 Montvale Avenue
Stoneham, MA 02180

10 9 8 7 6 5 4 3 2 1

Printed in the United States of America

Contents

Preface

Motives

Why write an introductory book on molecular cloning? One important consideration was the growing antiscience attitude among Americans, particularly young people. My response has been to write a book that, among other things, tries to show that recombinant DNA is not something to be mistrusted or feared. I want people to see genetic engineering as I see it: an exciting, intellectually stimulating enterprise; an area of study that is bound to accelerate our understanding of how living things work; a growing technology that will have a largely positive impact on our lives.

Many people are put off by genetic engineering because of its technical nature. To make recombinant DNA less intimidating, I thought it would be useful to generate a book that was somewhat less comprehensive and more accessible than some of the textbooks with which I was familiar. Such a book would certainly be within reach of most high school biology teachers. It might also be useful for middle and upper level managers at biotechnology companies – managers who aren't working at the bench but who may have to make decisions based on technological issues. At the same time, I thought a book aimed at this level would provide material for college survey courses.

I also wanted to write a book because I thought it would be enjoyable to combine writing with computing. Over the past ten years I've had a love affair with personal computers, and I've become convinced of their potential value as teaching and communication tools. My feeling at the time I undertook this project was that it might be enjoyable to use a computer to build a book, doing the layout and the illustrations, retaining considerable control over its appearance. As it turned out, control and responsibility are directly proportional. With increased control comes a newly required increase in attention to details. Even more worrisome, I had to make a multitude of difficult aesthetic decisions. In the end it was often enjoyable, but not always.

Finally, I was motivated by the thought that a book would represent something tangible, something solid that I could construct out of a half dozen years of ephemeral lectures and old floppy discs. It would be, I thought, something concrete that I could point to when asked what I'd contributed – aside from my research – to society over the last decade. Unlike lectures that vanish into the ether, a book is something that can be touched and put on a coffee table or sent to a mother in Florida.

Simplifying molecular biology

A few words about some of the difficulties involved in trying to "popularize" molecular biology. It seems to me that there are two great unifying principles of biology. The first is evolution. The second is that, apart from evolution, there are few great unifying principles in biology. Darwin has taught us that evolution proceeds by the sequential selection and fixation of a series of accidents. In effect, that means that organisms can use whatever means are available to solve a problem. In turn, this means that while there are general rules and even basic principles, exceptions abound.

It is because of these exceptions that biology in general, and molecular biology and genetic engineering in particular, are so rich in phenomena. That is also why biology texts are so long and dense. I have tried to shorten and lighten this book by consciously omitting mention of some of the exceptions and by trying to emphasize general principles. There are a few places, especially in Chapter 11, where I have examined some topics in more detail. But for the most part, if readers want a fuller treatment of any subject, they will have to look elsewhere. I have provided some references in the Bibliography to help those who have been stimulated (or frustrated) by this simplification. A computerized tutorial is also available for those who may want additional help in a nontraditional form.

Vocabulary

There is also the problem of vocabulary. I've included a Glossary at the end of the text. But glossaries are only stopgap measures because genetic engineers, like most biological scientists, seem to coin words almost as fast as the government prints money. This "word inflation" can best be appreciated by perusing a modern high school biology text. Apparently, learning biology has become an exercise in vocabulary building. I've tried to avoid this situation as best that I can by studiously avoiding the introduction of jargon and by using a descriptive phrase instead of a recently coined term. The disadvantage of this approach is that readers may come across unfamiliar terms in their peripheral reading. But I feel it is a small price to pay if it helps make the principles of genetic engineering and molecular biology more accessible to the general reader.

Organization

This book is divided into four sections. The first four chapters form an introduction to the foundations of molecular biology. Most people who have taken a biochemistry course in the last decade can safely skip this part. The second section (Chapters 5 through 10) deals directly with recombinant DNA technology, beginning with a short description of the essential techniques and organisms that are used in recombinant DNA, and continuing with an extended treatment of these topics. The third section is concerned with applications: how recombinant DNA has been used, and will be used, in the marketplace. It consists of a single chapter. The book ends with a section containing two chapters. One touches on the use of the computer in recombinant DNA technology. The other carries a brief discussion of the ethical considerations that underlie the use of genetic engineering.

Acknowledgments

I am very grateful for the many suggestions and criticisms that I've received over the years from students and colleagues to whom I've presented this material in lectures and seminars. I'm particularly indebted to Tim Stearns, Shahid Imran, and Steve Lawrence for examining the text and for providing feedback. Doctors Hubert Lechevalier and Carl Price gave encouragement at critical times. Finally, I am beholden to my wife, Gail, for her continuing support throughout this endeavor.

Introduction
to
Genetic Engineering

1

Large and small molecules

A fundamental concept of molecular biology, one that cannot be emphasized too strongly, is that the molecules of life fall comfortably into two categories: small and large.

Small molecules

Small molecules abound in living things. It is difficult to estimate their exact number, but there are certainly thousands of different ones in many cells. Most of the important ones have been isolated and purified, and their place in the vital scheme worked out. Some, for example, are burned as fuel – like the simple sugars and fatty acids. Others are the currency of energy for the cell, transferring energy generated by oxidation to the organelles and biomachinery that will use it. Still others are intermediates in metabolism, transients between the major biosynthetic and degradative steps in the living process. But, while small molecules are extremely important – even vital – they can largely be ignored in our effort to understand the basic principles of molecular biology. They can be ignored, that is, except as they relate to the other major class of molecules: the large ones.

Large molecules

Large molecules are small molecules strung together. Instead of synthesizing large biological molecules like small ones – by adding carbon, oxygen, hydrogen, and nitrogen atoms here and there, willy-nilly – Nature decided to make big molecules by simply linking together smaller ones in long linear chains. In other words, large molecules are **polymers** made up of many small molecules called **monomers**.

Proteins, DNA, and **RNA** are the major classes of large molecules (**macromolecules** is the technical term) studied by molecular biologists.

Proteins

Proteins are composed of assemblages of a class of small molecules called **amino acids** that are joined to one another in long unbranched chains. A depiction of the three-dimensional structure of one amino acid (**alanine**) is shown at the left. A more simplified view is shown in the illustrations below.

Notice that amino acids are composed of four parts (ignoring the H atom sticking out from the bottom).

First, there is the central carbon atom.

Attached to it, as shown in the illustration at the right, is a substituent called an **amino group**. The amino group has some of the character of the familiar household chemical, ammonia: In chemical terms it is basic (meaning that it behaves like a base; that is, it can neutralize an acid), and it is positively charged under most conditions in the cell.

Third, on the other end of all amino acids, is a **carboxyl group**. Carboxyl groups are found in such familiar substances as acetic acid (the main ingredient in vinegar) and they are responsible for the acidic character of amino acids. Carboxyl groups are negatively charged at neutral pH.

Finally, there is the **side group** (in the amino acid alanine, it is CH_3, as shown on the right). All the other substituents mentioned above are the same in virtually every amino acid. But the side groups differ and, in fact, are

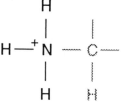

responsible for making each amino acid behave distinctively. For example, some of the side groups are oily; that is, they repel water. They tend to make the amino acid less soluble in aqueous solutions. Other side groups attract water and thereby increase an amino acid's water solubility. Still others are charged, either negatively or positively, and impart this charge to the corresponding amino acid.

Amino Acid	3 - Letter Name	1 - Letter Name
Glycine	GLY	G
Alanine	ALA	A
Valine	VAL	V
Leucine	LEU	L
Isoleucine	ILE	I
Serine	SER	S
Cysteine	CYS	C
Methionine	MET	M
Tyrosine	TYR	Y
Phenylalanine	PHE	F
Tryptophan	TRP	W
Histidine	HIS	H
Arginine	ARG	R
Lysine	LYS	K
Aspartic acid	ASP	D
Glutamic acid	GLU	E
Asparagine	ASN	N
Glutamine	GLN	Q
Proline	PRO	P
Threonine	THR	T

All in all, about 20 different amino acids are found in proteins. Their names are shown in the table at the right. Most proteins consist of a sequence of 100 or more amino acids, and usually each of the 20 amino acids is represented at least once, and often many times, in any given protein.

Protein formation

Proteins are formed when amino acids become linked together by strong (**covalent**) chemical bonds. Each protein originates when two amino acids react with one another. The carboxyl group of one reacts with the amino group of another – in a chemical reaction that eliminates water – to

loss of water

form a **peptide bond**. Notice that the product of the reaction – a dipeptide – retains one unreacted amino group at one end of the molecule (called the **amino terminal end**) and one carboxyl group at the other (called, obviously, the **carboxyl terminal end**). (In the example shown, the dipeptide consists of two units of alanine.). The free carboxyl group may react with the amino group of another amino acid, thereby forming a tripeptide. This process can be repeated over and over again. Asshown later, proteins are synthesized unidirectionally: New amino acids are always appended to the carboxyl terminal end of the growing peptide chain. Consequently, the amino terminal end of a developing peptide doesn't change during protein synthesis. Eventually, hundreds or even thousands of amino acids may be added to form a complete polypeptide chain. And again, to repeat by way of emphasis, any of the 20 amino acids may be added at any step during protein synthesis.

An aside about nomenclature: Because proteins are formed by a series of peptide bonds, they are sometimes called **polypeptides,** although, as noted below, there is a subtle distinction between a protein and a polypeptide.

And now a critical point: The character of a protein is determined by the sequence of its amino acids.

That is, it's not the number of kinds of amino acids in a protein or even the proportion of the various amino acids that make a particular protein distinctive. It is the *sequence* of amino acids, their order from one end to the other, that distinguishes one protein from another. For example, a very small protein might consist of ten amino acids: GLSQRSTEDI. Another might have the same amino acids arranged in a different order (LQSEIGSRTD, for example). These two proteins could be quite different from one another, each with distinctive physical and chemical properties.

A prediction

There is a considerable body of evidence supporting the statement made in the previous paragraph. Most of the data comes from experiments in which an amino acid substitution is made in a protein. If it is true that the order of amino acids in a protein is important, then changing that order should alter the character of the protein.

And that's what happens. Sometimes the result is very subtle. That is, the protein doesn't change very much. That's especially true if the amino acid that is substituted is very similar to the original. But in other instances, switching even a single amino acid for another can change the fundamental nature of the protein.

Sickle cell anemia provides a classic example. The protein hemoglobin is found in high concentration in the red blood cells of vertebrates. In essence, red blood cells are tiny bags full of hemoglobin that carry oxygen from the respiratory organs to the various cells of the body. Hemoglobin consists of four chains of amino acids: four polypeptides. There are two identical chains of 141 amino acids called α-globin and two β-globin chains of 146 amino acids. Individuals with sickle cell anemia are born with β-globin chains that contain a valine at amino acid #6 (numbering from the amino terminal end) instead of the glutamic acid that normally occurs there. This change (the result of a **mutation** – see Chapter 4) affects the structure and function of the hemoglobin molecule. In turn, the red blood cell itself becomes distorted, taking on the familiar sickle shape that characterizes the disease. These sickled cells get stuck in capillaries and eventually impede blood flow, causing further complications. The disease is so serious that without transfusions (and often despite them) people with the condition often die before becoming adults.

Some other characteristics of proteins

• **Proteins vary considerably in size.** They may consist of from tens to thousands of amino acids (an average number is about 350). The current holder of the record for the world's largest protein is **titin**, a polypeptide of some 25,000 amino acids found in vertebrate striated muscle.

• **Polypeptides may group together.** Up until now, the terms **proteins** and **polypeptides** have been used somewhat loosely and interchangeably. However there is a distinction between the two. A polypeptide consists of a single, unbranched chain of amino acids linked together by peptide bonds. In contrast, a protein can be a single peptide chain, or it may consist of two or more polypeptides in very close association with each other, as in the case of hemoglobin. These associations between chains, although intimate, do **not** occur by the formation of interchain peptide bonds. Instead, the proteins most often interact through, and are joined together by, a series of many weak bonds.

• **Peptide chains are irregularly shaped.** The illustration of the hemoglobin molecule shown on the right demonstrates another important point about polypeptides: Their chains are not straight rods. The different amino acids interact with one another so that the typical polypeptide bends back and forth, on top of and under itself, forming a complicated three-dimensional network. Protein structure and

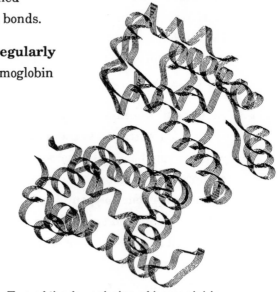

Two of the four chains of hemoglobin

how it relates to protein function are discussed in Chapter 2.

DNA

Deoxyribonucleic acid (DNA) is another polymer. The monomers from which it is formed are called **nucleotides**, or more precisely, **deoxyribonucleotides.**

There are four kinds of deoxyribonucleotides: **deoxyadenosine-5'-phosphate, deoxycytidine-5'-phosphate, deoxyguanosine-5'-phosphate,** and **deoxythymidine-5'-phosphate** (or A, C, G, and T). They all look rather alike, differing only in one component. They consist of three parts:

• A five-carbon sugar, pictured above, called **deoxyribose**. (Each carbon atom is numbered. The number five carbon is the topmost one on the left. Notice the absence of the OH group on the number 2 carbon atom.)

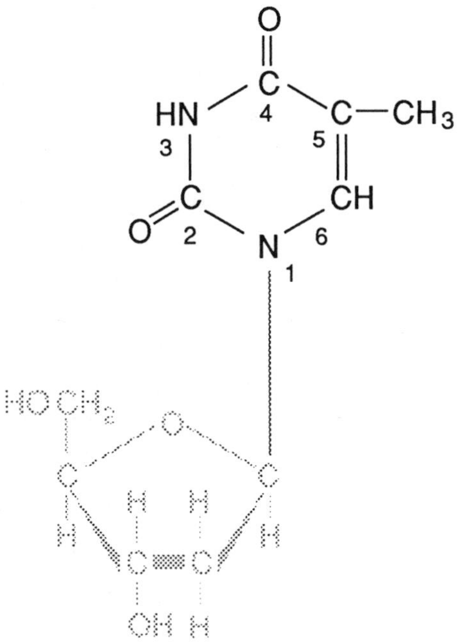

• A ring-shaped nitrogen-containing structure called a **base**. (Here is where the difference between the four different nucleotides lies. The four choices are **adenine, cytosine, guanine**, and **thymine**. The one shown at the left is thymine. Its numbering starts at the nitrogen atom that connects to the sugar.) Notice that the atoms of the base and sugar each have their own numbering system. To avoid confusion when they are both present in the same molecule, the numbers on the sugar are followed by the prime symbol, (').

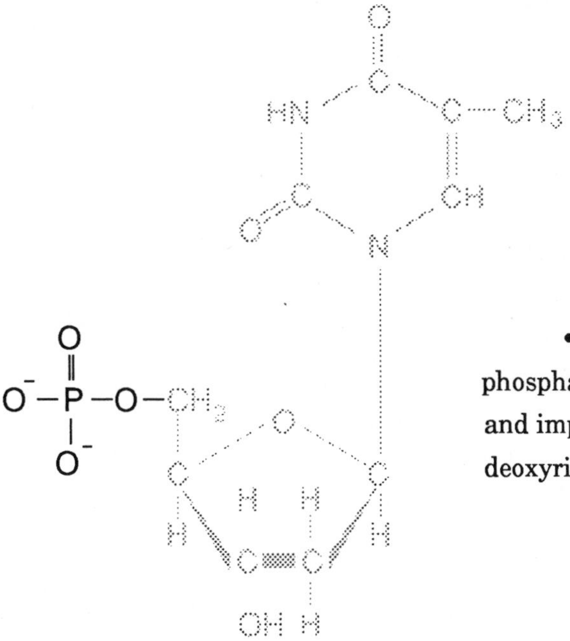

• A **phosphate group**. The phosphate group is negatively charged and imposes a negative charge on the deoxyribonucleotide and hence on DNA.

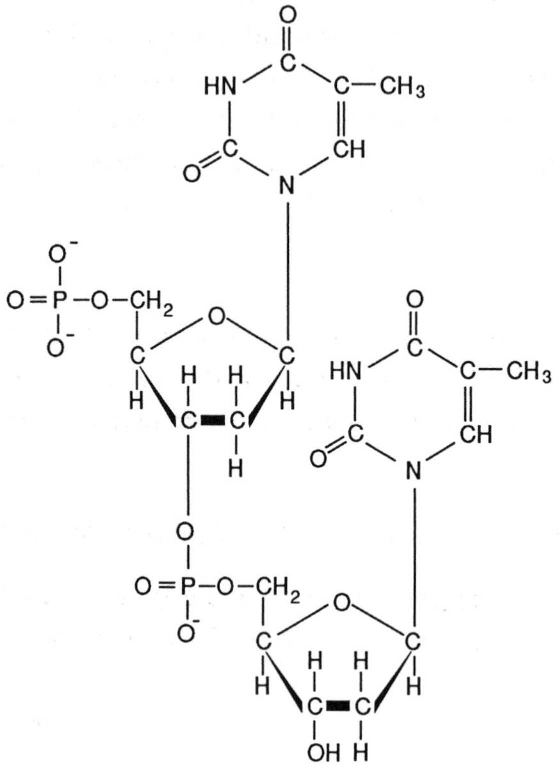

The deoxyribonucleotides are linked together via their phosphate groups by strong bonds forged like the peptide bond, by the elimination of water. A dinucleotide is shown at the left to illustrate the structure of the bond. The resultant polymer, which may contain hundreds of millions of nucleotides, is called DNA. Usually, DNA consists of two intertwining chains (or strands) that take the general shape of a helix (spring-shape). The structure of DNA is discussed in greater detail in Chapter 4.

RNA

Ribonucleic acid (RNA) greatly resembles DNA in that it also is composed of chains of nucleotides. But in the case of RNA, these are **ribonucleotides** rather

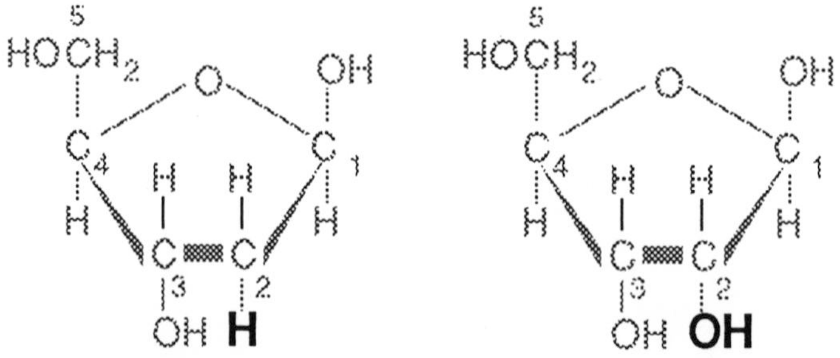

deoxyribose   ribose

than deoxyribonucleotides. It should be easy to guess that RNA contains ribose sugars instead of the deoxyribose sugars of DNA. (Ribose has an OH group attached to the second carbon instead of the H of deoxyribose.) Another difference between the two nucleic acids is that the base thymine (T) of DNA is replaced by **uracil** (U) in RNA. Also, in most instances RNA molecules are single stranded, in contrast with DNA, which is usually double stranded.

Summary

It's easy to lose sight of the main points because of the wealth of details that biology provides, and the flood of vocabulary in this first chapter may inundate the uninitiated. However, there are two cardinal principles that are important to take home.

There are two classes of molecules in living things: small and large.

The large molecules – the macromolecules – are polymers of a subset of small ones and come (for the purposes of this book) in three varieties: proteins, DNA, and RNA.

2

Proteins

The three major classes of macromolecules – proteins, DNA, and RNA – are all vital to living things. This chapter takes a look at proteins.

What do proteins do?

Proteins are, after water, the major constituent of cells. And, as befits their abundance, they are involved in virtually all cellular processes.

Some proteins function as structural components: the steel, lumber, and bricks of cells. Actin, one of the most abundant of all proteins in higher organisms, is one such molecule. In general, the presence of actin is associated with movement. In the illustration at the right, actin has formed several filaments that serve to help the cell move about. The contraction of our heart and the movement of our limbs are mediated by muscle cells, each of which contains a molecular motor that is composed of actin in combination with other muscle proteins.

The straight lines represent actin filaments coursing through a normal cell in tissue culture. The oval structure in the center of the cell is its nucleus.

Other proteins act as molecular police officers, recognizing and fastening onto undesirable elements, like bacteria and viruses, so that they can be eliminated from the organism. The **antibodies** of vertebrates are good examples of proteins that function in this capacity. Antibodies are discussed in greater detail in Chapters 9 and 11.

Still other proteins function to transport molecules into and out of cells. They serve as molecular pumps, transporting in some foodstuffs and desirable salts and forcing out waste products and poisons.

It should be obvious by now that proteins are extremely versatile, despite the fact that the functions so far enumerated are but a small fraction of what they can do. Yet, one of their properties is paramount: Proteins serve as specific **catalysts**, directing and accelerating the multitude of chemical reactions that occur in cells. As such, they help break down complex substances into simpler ones. They also aid in piecing together a variety of small molecules from parts of other little ones. And they play a crucial role in the synthesis of other proteins as well as DNA and RNA. The proteins that act in this way are called **enzymes**.

Enzymes

Enzymes direct chemical reactions in all living things by acting as biological catalysts. A catalyst speeds up the rate of a chemical reaction without being used up in the process. Without catalysts, most of the reactions that go on in living things would proceed far too slowly – by several orders of magnitude – to support life. Enzymes are protein catalysts, differing from familiar chemical catalysts like platinum and palladium in their extraordinary specificity. Generally, they only catalyze a single chemical reaction.

How much do enzymes speed up reactions?

Extraordinarily. In some cases they may accelerate the rate of a chemical step more than hundreds of trillions of times that of an uncatalyzed reaction. Most human-made catalysts are much less effective.

How much catalysis occurs in living cells?

There are thousands of different kinds of small molecules in a cell. Compare this to the estimated 2000 to 3000 different proteins that may be present. The vast majority of these are enzymes, and most of these affect the rate of a single chemical step. This means that hundreds or even thousands of different reactions are being catalyzed at any given time within a cell. It would appear that the problem of coordinating this frenzy would be enormous. And perhaps it is. Nevertheless, organisms are able to cope. The result is an orderly series of reactions, with one product often serving as a reactant for another. Each enzyme seems to play a unique role in the complicated pathways of chemical synthesis and breakdown that are going on, allowing the cell to function as an efficient factory.

How do enzymes work?

This question would require a long answer to summarize what is known and the large amount that has yet to be learned. However, in simplified terms, enzymes are thought to speed up the rate of reactions in three ways.

First, they bring chemical reactants (called **substrates**) very close together so that they have an increased opportunity to interact. Most chemical reactions occur when two or more molecules collide with each other. Enzymes seem to increase the local concentration of reactants by binding to them and dragging them into proximity.

Second, by binding to substrates in only certain orientations, enzymes align their reactants so that they interact more efficiently with each other.

Third, enzymes chemically react (reversibly) with one or more substances, forming intermediates that are more readily converted into the required products.

In order to carry out all these activities in a specific manner, enzymes must take on a particular three-dimensional structure. Expressed another way, enzyme catalysis requires that certain amino acids in the enzyme be positioned so that they can interact properly with the reactants. If this positioning is faulty or disrupted, the enzymes will no longer work.

How do enzymes get their three-dimensional structure?

The amino acids of proteins are able to rotate around the peptide bonds that join them, thereby allowing polypeptides to assume many different configurations. Yet enzymes have to assume a certain three dimensional structure so that they can operate suitably. How does an enzyme take on a specific shape?

There had been much debate about this question, but now it is widely accepted that the conformation of a protein is determined primarily by its amino acid sequence. In other words, a given sequence of amino acids will naturally and of its own accord fold in space to take on a specific three-dimensional shape. Apparently, many weak interactions occur among the different side chains of the amino acids that comprise the protein, and one particular conformation of the protein forms a uniquely stable structure.

A series of experiments were undertaken in an effort to test this idea. Two were particularly revealing.

First, researchers made active enzymes by chemically synthesizing a given amino acid sequence outside the confines of a cell. That is, they used a series of organic chemical reactions to knit together amino acids into a polymer with the same sequence as that produced by living cells. In most cases these artificial enzymes took on a specific shape, without benefit of help from any elements in the living cell. What's more, they worked properly in catalysis.

Second, an enzyme's native structure was disrupted (**denatured**) by agents (called, naturally enough, **denaturants**) such as acids, high temperatures, and detergents. A variety of tests indicated that the protein had lost its normal shape. It certainly had lost all detectable enzyme activity. Then the denaturing agents were slowly removed. After a while, the protein regained its original shape. At the same time, enzyme activity was restored.

These classic experiments offer good evidence that formation of a specific shape is inherent in a protein's amino acid sequence and that proper folding does not necessarily require the intervention of other molecules, such as other enzymes.

Summary

The main constituents of cells, after water, are proteins.

Proteins are versatile macromolecules, but their most important role is to act as enzymes – biochemical catalysts.

Enzymes differ from human-made catalysts in their great specificity and effectiveness. Without them, the chemical reactions in living beings would occur too slowly to support life.

Enzymes work by bringing the reactants of a chemical reaction into juxtaposition and by combining with them in ways that make their reaction energetically more favorable. For these properties, enzymes need to have specific three-dimensional structures.

The specific three-dimensional structure of an enzyme is stipulated by the sequence of its amino acids.

3

Protein synthesis

We have seen that the structure of an enzyme is critical to its function. We've also learned that an enzyme's structure is determined by its amino acid sequence. That brings up the next crucial question: How do organisms build proteins with the correct amino acids at each position? It would seem to be an incredibly difficult task to synthesize any large protein with a specific amino acid sequence. First, there is the problem of yield. Imagine a protein-synthesizing machine. Suppose that it added on amino acids one at a time with 90% efficiency at each step. A simple calculation will show that after only a few tens of amino acids, the efficiency of the total synthesis will be considerably below 1%. But cells do this kind of synthesis all the time, building as many as thousands of different proteins at any moment, at an efficiency of very close to 100%.

And that is only part of the problem. In a laboratory synthesis, a person or a computer tells the chemist which one of the 20 amino acids should be added at each position. Who or what tells the protein-synthesizing machinery of the cell to insert the right amino acid at the right position?

DNA is the answer

It is DNA that carries the instructions that inform the machinery of the cell which of the 20 possible amino acids belongs at any position in all of its protein chains. In this way, DNA is like an enormous cookbook, full of recipes for making thousands of enzymes.

And like a cookbook, it doesn't actually do anything. It's simply a program, a musical score, a data set. As we'll see, some other members of the macromolecular cast read the information and translate it (literally) into flesh and sinew.

How does DNA dictate which amino acid goes where?

It is the *sequence* of nucleotides in DNA that is responsible for informing the cell where each amino acid will be inserted in a protein. The nucleotide sequence of DNA is a language of sorts – a code – that is somehow translated into a protein sequence. Molecular biologists unraveled this code in the 1960s and their accomplishment represents one of the great triumphs of modern biochemistry and molecular genetics.

A genetic code

How is the code built into the DNA sequence decoded into an amino acid sequence?

Both DNA and proteins are linear (unbranched) polymers. As discussed above, any position in the DNA sequence may be occupied by one of four different nucleotides. Similarly, one of 20 different amino acids may be located at any given position in a protein. It should be obvious that a single nucleotide in DNA can't control the positioning of a single amino acid in a protein because there are too few kinds of nucleotides (or too many kinds of amino acids). Some kind of scheme is required where a group of nucleotides in DNA would direct the placement of specific amino acids. As shown in the figure, groups of two nucleotides can only stipulate 16 different amino acids. Groups of three

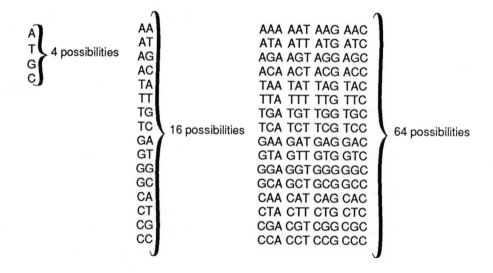

nucleotides are capable of specifying as many as 64 amino acids, but that seems wasteful since only 20 need to be coded for.

The triplet code

But, wasteful or not, three turns out to be the right number. That is, the DNA sequence is read in groups of three nucleotides, and each group of three nucleotides – no more, no less – specifies a given amino acid. This is the celebrated **triplet code**. For example, if a group of three consecutive thymine nucleotides is found in DNA, then the cell machinery knows to insert the amino acid phenylalanine in a particular position in the protein. If the three T's were to be followed by three A's, then a lysine would follow the phenylalanine in the protein. (The table to the right gives the complete code.)

A triplet code like that described above is capable of directing the insertion of up to 64 different amino acids in proteins. But since there are only 20 amino acids, it should be clear that either some groups of three nucleotides are not used for amino acids, or that some amino acids are coded for by more than one triplet.

First Position	Second Position				Third Position
	T	C	A	G	
T	PHE	SER	TYR	CYS	T
	PHE	SER	TYR	CYS	C
	LEU	SER	stop	stop	A
	LEU	SER	stop	TRP	G
C	LEU	PRO	HIS	ARG	T
	LEU	PRO	HIS	ARG	C
	LEU	PRO	GLN	ARG	A
	LEU	PRO	GLN	ARG	G
A	ILE	THR	ASN	SER	T
	ILE	THR	ASN	SER	C
	ILE	THR	LYS	ARG	A
	MET	THR	LYS	ARG	G
G	VAL	ALA	ASP	GLY	T
	VAL	ALA	ASP	GLY	C
	VAL	ALA	GLU	GLY	A
	VAL	ALA	GLU	GLY	G

Both things are true. Three of the triplets fail to specify any amino acid and in fact are signals to the protein that it is finished being synthesized. (These are called **termination triplets** and are indicated by the word "stop" in the table on the previous page.) Of the 61 remaining triplets, most are used more than once, some as many as six times.

Protein synthesis

The detailed biochemical mechanism by which proteins are synthesized is well understood, but it is complicated– too complicated to cover in detail here. A simplified diagram is shown on the next page, and the legend presents some additional information. But the major take-home lessons are these:

- DNA doesn't participate directly in protein synthesis.

- It acts through an intermediary.

- That intermediary is RNA.

In brief, DNA participates in protein synthesis in the following way. First, DNA in its familiar double-stranded state makes a copy of a part of one of its chains in the form of RNA (a process that is described in greater detail in the next chapter). In turn, this RNA (called **messenger RNA or mRNA**) enters a very large and complex subcellular molecular machine called a **ribosome** (see the diagram on the next page), where its sequence of nucleotides is directly translated into the sequence of the protein. The correct insertion of any given amino acid is determined by the genetic code that was mentioned above.

Translation is actually a technical term. It comes from the fact that the protein sequence is written in one language (that of the 20 amino acids) and the DNA and RNA sequences are written in another (that of the four nucleotides). The manufacture of RNA from DNA isn't translation because both are written in the language of nucleotides. DNA-directed RNA synthesis is therefore called **transcription,** because it involves the rewriting of text in the same language but in a different form (as between longhand and typewritten text).

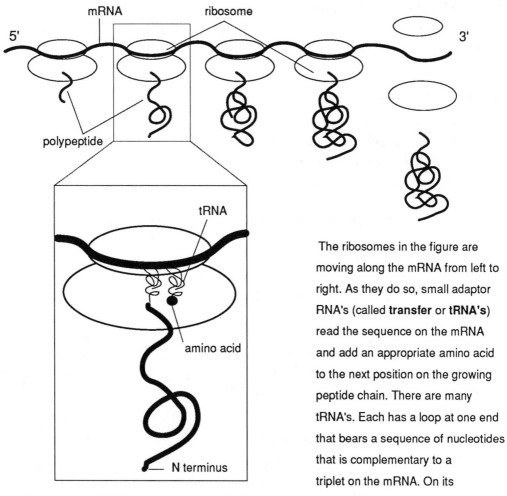

The ribosomes in the figure are moving along the mRNA from left to right. As they do so, small adaptor RNA's (called **transfer** or **tRNA's**) read the sequence on the mRNA and add an appropriate amino acid to the next position on the growing peptide chain. There are many tRNA's. Each has a loop at one end that bears a sequence of nucleotides that is complementary to a triplet on the mRNA. On its other end, the tRNA bears an amino acid. The particular amino acid that the tRNA carries corresponds to the sequence of the complementary loop near its other end. Notice that the protein is synthesized beginning at its amino terminal end (called the **N terminus**).

Summary

DNA specifies the position of all 20 amino acids in proteins.

It does so by utilizing a triplet code.

DNA acts indirectly – through mRNA – to translate the language of nucleotides into the language of proteins.

4

More about DNA

DNA, as already mentioned, consists of nucleotides – deoxyribonucleotides – linked together in long linear arrays. One chain is shown below. It is only a tetranucleotide, but it will serve for illustrative purposes. Notice the nature of the bond that links adjoining nucleotides. A phosphate group (the bold faced P in the figure) connects carbon 5 of the sugar (the 5' carbon – the prime distinguishes this carbon from the fifth carbon in the base) to carbon 3 of the sugar of the next nucleotide. Notice that this imparts an asymmetry to the molecule. As indicated in the figure, there's an unattached phosphate group on the 5' carbon of a deoxyribose at one end of the molecule (called the **5' end**, it is at the top of the figure) and a free OH group on the 3' carbon of a deoxyribose at the other end (called the **3' end**).

Double-stranded DNA

Drawn this way, the structure of DNA doesn't seem suggestive of its purpose. However, when viewed three dimensionally and in its double-stranded form, the DNA takes on a familiar and more informative appearance. Typically, DNA comes in two strands with the bases of each strand pointing inward like the rungs of a ladder, and the sugar-phosphate linkages of each strand playing the role of the limbs of the ladder. But it is a strange ladder in that the two limbs twist around each other like two intertwined springs or helixes. One strand goes in one direction (5' to 3' from top to bottom); the other strand goes in the opposite direction (3' to 5' from top to bottom). Because of the proximity of the two strands, only two combinations of bases are possible at any given position. In particular, if one strand has an A occupying a certain position, the corresponding base on the other strand must contain a T. Similarly, if one strand has a G in one position, the other strand must have a C opposite. The A/T and G/C combinations on apposing strands are called **base pairs**, and two molecules of DNA that have these combinations on opposite strands are said to be **complementary.** In an exactly analogous manner, single-stranded RNA can form base pairs with single-stranded DNA, except that A/U base pairs substitute for A/T ones.

Base pairing

Certain conditions, for example high temperatures, will disrupt the pairing of the two complementary strands of DNA (or complementary strands of DNA and RNA), and the strands will separate – a process called **denaturation** (analogous, but not exactly equivalent to protein denaturation discussed in Chapter 2). If temperatures are moderated and a solution of denatured DNA is allowed sufficient time, each strand will tend to find its complement and reform the original double-stranded molecule again – a process called, naturally enough, **renaturation** (or more formally **nucleic acid hybridization**).

The important point to remember about nucleic acid renaturation is that it shows that complementary strands of DNA have an affinity for each other. As we shall see, Nature and genetic engineers have put nucleic acid hybridization to use in many different ways.

DNA replication

Among other things, the double-stranded structure of DNA immediately suggested to James Watson and Francis Crick (who first proposed the double helix structure in 1953) a mechanism by which the molecule could reproduce (the technical term is **replicate**) itself. A simplified picture of a replicating molecule is shown on the facing page.

The replication of the DNA of even the simplest organism is a complicated process. An assortment of enzymes, proteins, and many steps are involved. The details differ from organism to organism. There are, however, several basic rules that govern DNA replication in all known creatures.

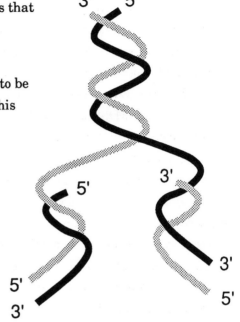

First, DNA replication is said to be **semi-conservative**. In plain words this means that each new double-stranded daughter molecule consists of one intact strand that comes from the parent DNA and one strand that is newly synthesized (see the figure above).

Second, synthesis of the new strand always proceeds by the addition of nucleotides on to the 3' end of the growing chain.

Third, the major step in DNA synthesis is mediated by an enzyme called DNA polymerase. This enzyme takes each of the four nucleotides and adds them stepwise to the end of the newly forming chain.

Fourth, DNA polymerase 'knows' which nucleotide to put at any given position because it is using one strand as a **template**. The template is a single-stranded molecule of DNA (the intact strand mentioned in rule 1) which, by base pairing, determines the sequence of the newly synthesized second strand. As synthesis proceeds, the enzyme puts an A across from the template's T; a G across from the template's C; a T across from the template's A; and a C across from the template's G.

Fifth, and last, DNA polymerase needs a pre-existing chain to add nucleotides to. That is, the enzyme can't begin synthesis on its own (Readers who are puzzled by how replication actually gets started in light of this limitation of the enzyme will have to refer to one of the references cited at the end of the book). The required pre-existing chain is called a **primer**. It is single stranded and is complementary to a part of the template. In the presence of primer and template, DNA polymerase adds on nucleotide units to the primer in a sequence dictated by the template.

Transcription

The structure of DNA also suggests a mechanism for transcription - a process which is pictured at the right. In transcription the two strands of DNA separate temporarily so that one strand can be used as a template for the synthesis of a strand of RNA. The synthesis of the RNA goes from its 5' end toward its 3' end (like replication). Each nucleotide added is the complement of the base on the DNA. That is, at a position where the DNA has an A, the RNA adds a U; where DNA has a T, RNA has an A; and so on. Synthesis is directed by an enzyme called **RNA polymerase**. The enzyme starts RNA synthesis at certain well defined points on the DNA and ends at specific sites. Depending on the location and number of starting points, thousands of different RNA molecules can be synthesized from the DNA. Notice that transcription differs from replication in that only one strand of RNA is produced from the DNA. In addition, a ribonucleotide chain is produced, not two deoxyribonucleotide chains. There's one more difference between replication and transcription: transcription doesn't require a primer.

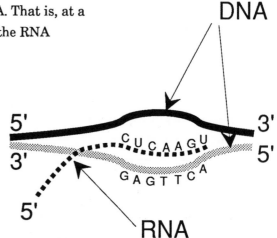

Some additional questions

The replication of DNA and the transcription of RNA are complicated processes. I've skipped over many of the significant details of each in order not to get us mired in minutiae. There remain, however, several important questions still to be dealt with and a few terms to be clarified.

What is a chromosome and how does it relate to what we've covered so far?

DNA is found in the nucleus of cells in the form of a few extremely long molecules. Each of these molecules (along with several other components) is called a **chromosome** and each contains the information for constructing many different proteins. If you remember your high school genetics, you'll recall that the chromosomes of most 'higher' organisms come in pairs. In most organisms, the mother and father each contribute one chromosome to each set of the offspring. Both chromosomes of each set usually carry the same information and are said to be homologs. That is, there is a two-fold redundancy built into the system. The advantage of doing things this way is best understood if one chromosome carries an error (as for example in the DNA sequence stipulating the β chain of hemoglobin mentioned above). Often, but not always, the presence of one homolog carrying a correct copy of a sequence will help remedy a mistake on the other. That is, the organism will appear normal, even though one of its genes is defective.

Each organism has a characteristic number of chromosome pairs. For example, *Drosophila melanogaster* – the little fruit fly famous for its use in genetic experiments – has four pairs of chromosomes, and each chromosome has sufficient DNA to code for thousands of proteins. Humans have 23 pairs of chromosomes. Again, each pair carries a different subset of genetic information. In combination, the nucleotide sequence of DNA in these chromosomes (in fruit flies, humans, asparagus, elephants or any other organism) carries all the information required to specify the amino acid sequence of all the proteins made by that organism in its lifetime.

What is a gene?

In humans there are about 3,000,000,000 (3 billion) base pairs worth of DNA in their 23 chromosomes. Surprisingly, only a portion of this vast amount of chromosomal DNA is transcribed into RNA. These transcribed areas are scattered throughout all the chromosomes. Each RNA molecule that is manufactured, therefore, represents only a tiny fraction of the total DNA. Although there are exceptions, each RNA molecule generally contains the information for making a single specific protein.

Armed with this information, we can define a gene in molecular terms. A gene is a section of a chromosome that contains the information to make a specific protein, through the production of a specific RNA.

However, a gene is also more than simply a transcribed portion of the DNA of the chromosome. It also commonly contains additional instructions, data encoded in the DNA sequence outside of the transcribed areas, that are responsible for 'telling' the cell how much RNA to make and in which tissues and under what circumstances to transcribe a given segment of DNA.

These controlling or regulatory regions of the DNA may act over long distances to modulate the genetic activity of the DNA, and they are an essential part of each gene. To a great degree, we don't understand the mechanism by which this control is exerted. When biochemists and molecular biologists figured out how the language of proteins derived from the language of nucleic acids, they were said to have cracked the genetic code. Nowadays, we talk about cracking *a* genetic code, because a large effort is underway aimed at understanding how the non-transcribed regions of the DNA actually control gene expression. Trying to figure out what the nontranslated part of the DNA is doing promises to be a larger undertaking than breaking the triplet code.

Summary

We've now completed our brief (and of course, incomplete) introduction to molecular biology. The following are important points to remember:

Living things have molecules that come in two flavors : small and large.

The large molecules are polymers composed of certain classes of the little ones.

The major classes of large molecules are proteins, DNA and RNA.

Proteins are made of small molecules called amino acids; DNA and RNA are made of small molecules called nucleotides.

Enzymes are proteins that act as specific catalysts in the cell. They do this by virtue of their three dimensional shape which derives directly from their sequence of amino acids.

The order of amino acids in a protein is dictated, indirectly, by the order of nucleotides in DNA.

A portion of DNA, a gene, transcribes an RNA molecule which encodes the sequence of a newly manufactured protein in a process called translation. Three nucleotides in the RNA specify one amino acid in the protein.

Finally, we've seen that DNA is double stranded and that the two strands are complementary; adenine on one strand is always matched by thymine on the other; guanine on one strand is matched by cytosine on the other.

5

Introduction to genetic engineering

Genetic engineering arose because biologists wanted to understand how genes are organized and how they work. By the 1970s, several generations of biologists had contributed to our understanding of these issues, but the story still seemed woefully incomplete. Part of the problem was that most of our information had come from geneticists. Like Mendel, who counted the offspring of pea plants, geneticists gain understanding by inference from the kinds and numbers of progeny that result from a cross. Most of the genetic data that had been collected over the years was suggestive and provocative (often based on beautiful and imaginative experiments), but they needed confirmation by complementary biochemical studies. However, in order to do biochemical experiments with genes, biochemists needed the means to purify them. Unfortunately, for the most part, purified genes were very hard to obtain.

Genes are chemically similar

It is important to understand why biochemists were finding it so difficult to purify specific genes, even as recently as the 1970s. First of all, in chemical terms DNA is a relatively simple polymer. As we have seen, it consists of only four kinds of units strung together in a linear array. And all four nucleotides are chemically rather similar. After polymerization, what results are long macromolecules that have nearly identical chemical properties regardless of their base composition, making different DNA's almost impossible to separate on the basis of their chemical behavior. Contrast this with proteins, the other large polymers that we've encountered. They are composed of 20 different kinds of units (amino acids) many with quite different chemical properties. Of course, they impart these properties to the proteins that they compose. Different proteins can be separated from one another (and have been for decades) because of their distinct chemical and physical behaviors.

Large pieces of DNA are very fragile

Then there is the problem of size. DNA – at least most of the nuclear DNA in higher organisms – consists of enormously long molecules. In fact, there is good evidence that each chromosome consists of a single molecule of double-stranded DNA. These large molecules are extremely fragile when removed from the cell. Simply drawing them up in a pipet or shaking the container in which they are collected will break most of them into many smaller pieces. And the DNA breaks randomly, not conveniently at the ends of genes.

Consequently, when biochemists extracted DNA from higher organisms, they were left with a jumble of randomly broken molecules of different sizes, all of which had similar chemical properties. Some of the pieces contained whole genes, some assemblages of more than one gene, and some broken genes. No wonder biochemists were dismayed.

The three steps of gene cloning

Gene cloning was invented in an effort to remedy this situation. Key to its understanding is that instead of using chemical or physical techniques to isolate genes, biological methods are employed. There are three major steps in the process.

First, the DNA of the organism containing the gene (or any sequence of nucleotides) of interest must be broken into smaller pieces that are convenient to work with. These pieces of DNA, some or all of which are the objects of the cloning procedure, are called **passenger DNA**.

Break DNA

Second, the pieces of passenger DNA must be joined (*in vitro*) to a second piece of DNA that can replicate itself and any attached passenger. This second DNA is often called a **vector** or **cloning vehicle**.

The result of the joining is a hybrid molecule, a **chimera**, a **recombinant DNA**, consisting of two kinds of DNA coupled to one another on a single piece (often a closed ring) of nucleic acid.

Vector

Cut open vector

Passenger

+

Third, the joined passenger and vector must be introduced into a living cell. The cell serves as a biological copying machine, making many exact copies of the recombinant molecule.

The process is called **molecular cloning** because the hybrid DNA replicates within the living cell without further recombination. In addition, the cell in which the

Hybrid DNA

recombinant molecule finds itself multiplies many times forming a colony. Each member of the colony is an exact duplicate of every other, and each carries many molecules of the recombinant DNA.

One additional step

In one sense, once a colony is generated with a hybrid DNA contained within it, the essence of molecular cloning has been accomplished. All the genetic engineer has to do is break open the cells, isolate the recombinant DNA, and work with it. In another sense, however, the work has just begun. That's because a typical genetic engineering experiment produces many colonies of cells, each of which may harbor a different passenger/vector complex. Thus the last step of molecular cloning: The identification of the clone that carries the particular passenger of interest.

How can recombinant molecules be used?

In the following chapters each of the steps of molecular cloning is described in more detail. But before forging ahead, I am going to discuss briefly how recombinant DNA molecules can be utilized. More specific examples are presented in later chapters, particularly Chapter 11.

• **The passenger DNA can be sequenced.** That is, a researcher can figure out the order of the successive bases in a piece of DNA. Determining the sequence of thousands or even tens of thousands of bases is now carried out, more or less routinely, in even the smallest of molecular biology and genetic engineering laboratories. In the near future, with the advent of automation and new techniques, the rapidity of DNA sequencing will increase by an order of magnitude or more. The human genome sequencing initiative proposes to take advantage of these advances to determine the DNA sequence of the three billion or so bases of the 23 human chromosomes. Besides new biochemical techniques, very large sequencing projects will require sophisticated computer programs to keep the enormous amount of data in a usable form. I discuss the current impact of computers on genetic engineering in Chapter 12.

• **The sequence of any protein can be changed.** In principle, any enzyme – or for that matter, any protein – can be altered by modifying the DNA that codes for it. Prior to recombinant DNA technology, proteins could be modified by chemical treatment, but that was often done with great difficulty, in low yield, and the range of changes that could be introduced was small. Now, virtually any aspect of a protein's primary sequence can be easily refashioned as long as the gene that codes for that protein has been cloned. By making these changes, genetic engineers dream of producing enzymes that can catalyze reactions that are unknown in nature or modifying enzymes to make them more stable or more efficient. Unfortunately, not enough is understood about the relationship of the structure of proteins to their function to take full advantage of the changes that the technology allows.

• **The function of genes can be deduced.** Molecular cloning has had an enormous impact on this area. The usual method of analysis is to change the DNA sequence of a particular gene and to determine what effect the change has on gene function. By this means, a scientist can reason backward and figure out what any given sequence does. In particular, alterations in DNA combined with transformation (the ability to introduce DNA molecules into living organisms) has already yielded exciting insights into how genes are regulated and controlled.

• **The passenger DNA can be used to synthesize a protein of interest, often in large quantities.** The interleukins, blood clotting factors, growth hormones, and growth factors are among the proteins that have made their way from the recombinant technology workbench to consideration in the corporate boardroom. This area has been the most visible manifestation of genetic engineering, and many more products are in the pipeline.

• **A gene can be introduced into an organism that is lacking or deficient in a particular function.** The old dream of gene therapy has become real. On the farm, transgenic animals and plants are playing an increasingly important role. For humans, whether it will prove to be a widely used tool or even one that will be practical in the near future is another matter. If gene therapy can be done in humans, we will have to wrestle with the ethical problems that this ability raises, especially if we mean to tamper with genes in the germ line. Some of these ethical problems are discussed in Chapter 13.

• **A cloned DNA sequence can be used as a diagnostic tool.** Prospective parents can be checked for a wide range of genetic disturbances. Embryos can be assayed for the absence or presence of genetic disease and undesirable traits. Paternity, maternity, and genetic relatedness can be established with unprecedented assurance. In the crime laboratory, DNA "fingerprinting" may complement the more familiar kind. The ethical and social problems that these techniques will introduce are enormous.

Summary

The first step in molecular cloning is to cut the passenger DNA into pieces of convenient size.

The passenger is then joined to a vector DNA, forming a hybrid (or recombinant DNA) molecule.

The recombinant DNA is then introduced into a convenient host.

Finally, the clone that harbors the desired recombinant DNA molecule must be identified.

6

Cutting and measuring DNA

The first step in gene cloning is cutting DNA into appropriately sized pieces. But how is the size of DNA molecules measured?

Agarose gels

The most popular way is via agarose gel electrophoresis. Agarose is a derivative of agar – an edible polysaccharide extracted from seaweed. When cooled from heated solutions, it forms a nearly transparent gel that looks like a slab of gelatin dessert. In the laboratory, suspensions of agarose in buffer are heated to boiling (often in microwave ovens, which have become fixtures in recombinant DNA laboratories), and the hot solution is poured on plastic or glass plates to a height of a few millimeters or so. After the gel sets, DNA solutions are placed in little depressions formed by leaving a comb in the gel. The gel is then flooded with a weak salt solution, and a voltage is applied. Since DNA is negatively charged (due to its phosphate groups), it moves toward the positive pole.

Agarose gel

Comb

Large DNA molecules advance slowly in the gel, probably because they are impeded by the gel matrix. Smaller ones encounter fewer barriers and move more quickly. By plotting the distance migrated against the reciprocal of size (in base pairs) of a group of DNA standards, a fairly straight line can be obtained. From these data, the length of unknowns can be easily calculated. For ease of comparison, the standards are usually run in the same gel beside the unknowns.

Gels of differing porosities can be made by adjusting the concentration of agarose. With 2 and 3% solutions, double-stranded DNA's as small as 50 or 100 base pairs can be resolved. More dilute gels can resolve fragments as large as about 30,000 base pairs (30 kilobase pairs or 30kb).

Polyacrylamide gels

When very small DNA molecules need to be analyzed or when high-resolution separations are required, electrophoresis is carried out in polyacrylamide gels. They, like the gels made from agarose, are water white. Unlike agarose, polyacrylamide gels are formed by the chemical polymerization of acrylamide, a small synthetic organic compound. Polyacrylamide gel electrophoresis is most often used to separate fragments of DNA that are between 6 and a few thousand base pairs in length. The technique is used mostly in DNA sequencing, a topic that is described in Chapter 10.

Detecting DNA

DNA doesn't have any color. How is it visualized in a clear gel after electrophoresis? DNA can be detected in both agarose and polyacrylamide gels after staining with **ethidium bromide**, a chemical that forms a fluorescent complex when it binds to DNA. Usually, the ethidium bromide is added to the gels before electrophoresis. After the run, the gels are examined and often photographed under an ultraviolet light. Yellow-orange zones (bands) of fluorescence appear, indicating migration of discrete pieces of DNA. Other agents –

Ethidium bromide

like methylene blue or silver stains — can also be used to visualize DNA, but they are less often used.

Separating large fragments of DNA

Very large fragments of DNA move readily in agarose gels when an electrical field is applied. But they don't separate according to size. In fact, very big pieces of DNA all migrate at about the same velocity regardless of whether they are 20kb in length or 2000kb.

If it becomes necessary to resolve very large molecules of DNA, even pieces the size of some chromosomes, several newly developed electrophoretic techniques can be used. In one of these, called **pulsed field electrophoresis (PFE)**, DNA molecules are analyzed on agarose gels but with a modified apparatus. Instead of having only two sets of electrodes situated at opposite ends of the electrophoresis device, the pulsed field apparatus bears four sets of electrodes. As shown in the figure, two are designated A- and A+ and the other two are called B- and B+. The two A's and the two B's are set at an angle of 120° with respect to one another. During electrophoresis, the DNA molecules are subjected to alternating bursts of current from the two A and B pairs, thereby pulling the DNA alternately to the right and to the left as it advances (from south to north in this case) through the gel. The DNA seems to reorient itself each time the current is switched, and the time that it takes for reorientation appears to be dependent on its length. The end result is an ability to separate very large pieces of DNA on the basis of their size, and a concomitant ability to estimate the length of an unknown when run beside standards of known size.

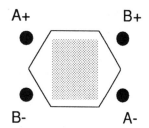

Another procedure, **field inversion electrophoresis**, has also recently been devised. In it, the electrodes are set up as in ordinary electrophoresis, but the current is switched so that the DNA first moves in one direction and then at a 180° angle in the opposite direction. Of course, if the switching were to be done in equal time intervals – say, 1 second forward and 1 second back – the DNA wouldn't move at all. To get net migration electrophoresis is conducted forward for a longer time than backward. One might expect that moving ahead three steps and back two would be equivalent to moving forward one step at a time, but that's not what happens. Apparently the DNA reorients itself during the switching cycles, as in pulsed field electrophoresis, and the rate of reorientation must differ for pieces of DNA of different sizes.

Restriction endonucleases

One critically important advance that has greatly stimulated the rapid progress in molecular biology and genetic engineering was the discovery of a set of enzymes that are capable of cutting DNA at defined sequences. These enzymes are found in a variety of microorganisms and are called **restriction endonucleases**, or more simply, restriction enzymes. The first specific restriction enzyme was discovered by Hamilton Smith in 1970, an accomplishment for which he was awarded the Nobel Prize. Since then more than 350 similar enzymes have been reported, many of which have different specificities.

Nomenclature

Smith and another Nobel laureate, Daniel Nathans, devised a nomenclature for these enzymes. In brief, the name of each restriction enzyme derives from the organism from which it is isolated. The first letter of the genus name plus the first two letters of the species name form the first three letters. If necessary, a letter indicating strain designation is added, and finally a number is appended that stands for the order in which the enzyme was discovered in each organism. For example, *Hin*dIII is the name of an enzyme that is isolated from the bacterium *Haemophilus influenza* strain d, and it was, presumably, the third restriction endonuclease identified from that source. The first three letters of the name are italicized.

The restriction enzymes owe their usefulness to the fact that they bind to DNA at specific DNA sequences called **recognition sites**. Almost all interact with a four, five, six, or eight base pair sequence before they cut the DNA at or near their recognition site. It should be obvious that enzymes that have a six base pair recognition site will, on the average, produce larger pieces of DNA than those that recognize four base pairs. Expressed quantitatively, the approximate size of the fragments produced by a particular enzyme, given that it is cutting a DNA containing an equal proportion of all four nucleotides, can be calculated from the formula:

$$\text{average size of fragment} = 4^N$$

where N is the number of bases that the enzyme recognizes. Hence, a four-cutter (the shorthand name for an enzyme that recognizes a site containing four base pairs) is expected to cleave random DNA into fragments of about 256 base pairs while an enzyme with a recognition site of six bases will produce pieces (on the average) of about 4096 base pairs.

Properties of restriction enzymes

Ends

As illustrated, restriction enzymes invariably cut DNA in such a way as to leave a 3' hydroxyl on one end, and a 5' phosphate on the other.

In addition, most (but not all) enzymes recognize a symmetrical site (see the figures on the next page).

Restriction enzymes break the bonds between nucleotides here, and generating an OH group on the 3' end of one fragment and a phosphate group on the 5' end of the other

Another interesting property of restriction enzymes is that while they often recognize a symmetrical site, they do not always cut at the axis of symmetry. For instance, the enzyme *Bam*HI (from *Bacillus amyloliquifaciens* strain H and pronounced BAM) recognizes the site GGATCC and cuts the DNA between the G's in a manner depicted at the right.

Note that the cut produces an overhanging 5' single-stranded end of four nucleotides on each of the two pieces that are newly liberated. Notice also that these two ends are complementary to one another.

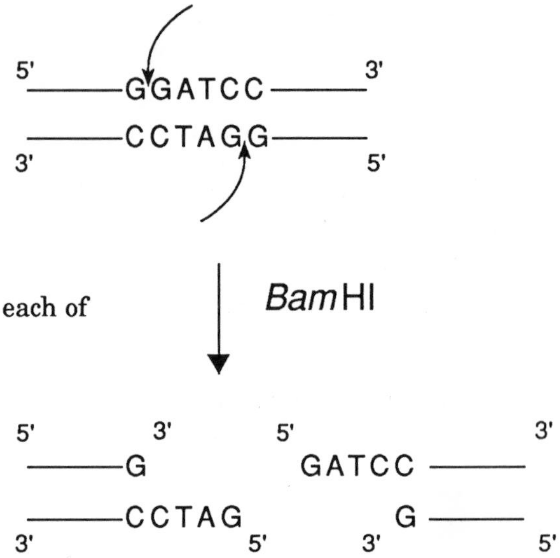

5'——G̓GATCC——3'
——CCTAGG̓——
3'   5'

↓ *Bam*HI

5'   3'   5'   3'
——G   GATCC ——
——CCTAG   G ——
3'   5'   3'   5'

Similarly, the enzyme *Bgl*II (from *Bacillus globiggi* and universally and irreverently pronounced BAGEL TWO) recognizes the sequence AGATCT and cuts between the first A and G residues.

Notice that *Bgl*II also yields a single-stranded 5' end. In fact, the four nucleotide single-stranded ends are the same for both *Bgl*II and *Bam*HI.

Moreover, there are at least two other six cutting enzymes that have been discovered that leave the same four nucleotide overhang: *Bcl*I and *Xho*II. These overhanging ends are very useful because —

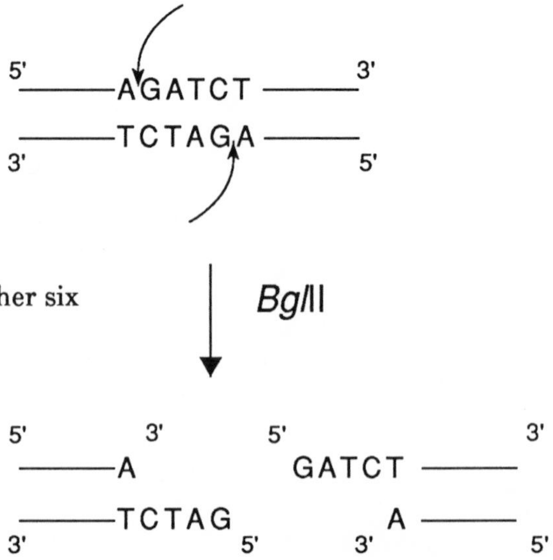

5'——A̓GATCT ——3'
——TCTAGA̓ ——
3'   5'

↓ *Bgl*II

5'   3'   5'   3'
——A   GATCT ——
——TCTAG   A ——
3'   5'   3'   5'

under the proper conditions – they may base pair with each other. In fact,
because of their affinity for one another, they are often called cohesive or **sticky
ends**. Moreover, if molecules with these ends are treated with the appropriate
enzyme – **DNA ligase** – their phosphodiester bonds may be rejoined (ligated).
When two ends that originate from digestion by a single enzyme are ligated, the
resulting molecule can be cut by the same enzyme again. But if the ends of a
DNA molecule that originated with a *Bam*HI cut and a *Bgl*II cut are joined
together, the new sequence will not be cut with either enzyme (See if you can
work out the reason why for yourself).

Note that all restriction
endonucleases do not generate
5' single-strand overhangs. In
fact, some don't even produce
an overhang at all. Several
enzymes – like *Sac*I – produce

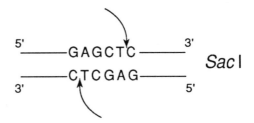

3' single-stranded sticky ends. And some enzymes – like *Pvu*II – cut at the axis of
symmetry, leaving perfectly aligned ends. DNA molecules without overhangs are
said to have **blunt ends**.

In addition there are restriction
enzymes that cleave DNA some
distance away from the sequence
that they recognize. For example
the enzyme *Hga*I makes staggered

cuts that lie 5 and 10 nucleotides away from a 5 base pair sequence, GACGC.
This leaves 5' overhanging ends, but, in contrast to the enzymes described above,
these will be different almost every time the enzyme cuts.

The utility of the restriction enzymes

The discovery of these many restriction endonucleases have allowed genetic engineers to cut pieces of DNA at specific sites and into defined sizes. The result has been that a scientist can work with a collection of molecules all of the same size and with ends of known sequence. Restriction enzymes have proved to be valuable analytical and diagnostic tools as well, as we will see in Chapter 11.

The utility of the restriction enzymes – to organisms

Why do many organisms carry enzymes that cut DNA, especially in view of the importance of DNA as the genetic material? The answer seems to be that the restriction endonucleases protect against invasion by foreign DNA's. They recognize these intruders (mostly viral in origin) and carve them into pieces.

But how does the organism's own DNA avoid this fate? Restriction endonucleases come paired with another set of enzymes called **modification enzymes**. These have the same recognition sites as their partner endonucleases. But instead of cutting the DNA, modification enzymes add a chemical group (a methyl group) onto one of the deoxyribonucleotides at the recognition site. (The modification enzyme pictured

here is *Eco*RI
methylase. It
has added a
methyl group to
the second A of
the site
recognized by
the restriction
enzyme, *Eco*RI.).
This
modification
prevents the
DNA from being
digested by
*Eco*RI.

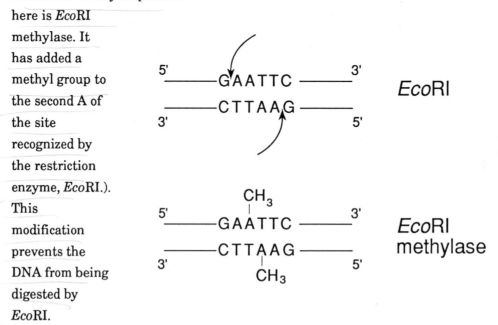

Ligation

Almost as important as the discovery of the restriction endonucleases was the uncovering of an enzyme activity (briefly mentioned above) that allows the joining of free ends of DNA segments. Two different **ligases** are known, but the one that comes from the bacteriophage T4 is most often used by molecular biologists for carrying out this reaction. It works best on sticky ends, but it will also knit together blunt-ended DNA if the DNA is in high enough concentration and if enough enzyme is added. In either case, for the enzyme to act properly, the molecules to be joined must have a phosphate group at their 5' ends and a hydroxyl at their 3'ends. As you may recall, these are precisely the kinds of ends left after restriction endonuclease scission.

There are three fundamental ways of joining DNA molecules together: sticky-end ligation, complementary-homopolymer ligation, and blunt-end ligation.

Sticky end ligation

Sticky ends are complementary single-stranded regions found at the ends of some DNA molecules. As mentioned above, many but not all restriction enzymes form single-stranded ends like this when they cut DNA. For example, when the restriction endonuclease *EcoRI* cleaves the sequence above, it leaves the ends at the right.

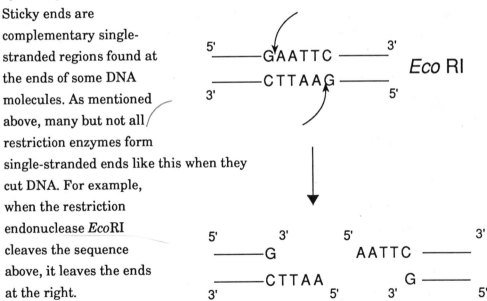

Under the appropriate conditions, in the presence of T4 DNA ligase, these ends can be tied together to reform a complete molecule.

Let's examine a specific case often encountered by the genetic engineer. If a circular DNA molecule (in subsequent chapters we'll see that many of the molecules that are used in molecular engineering are don't have ends) is cleaved once with *Eco*RI, a

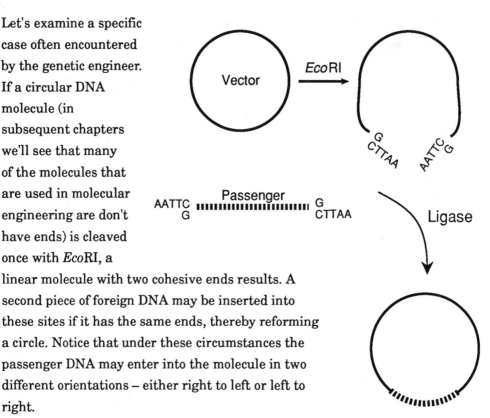

linear molecule with two cohesive ends results. A second piece of foreign DNA may be inserted into these sites if it has the same ends, thereby reforming a circle. Notice that under these circumstances the passenger DNA may enter into the molecule in two different orientations – either right to left or left to right.

Once the foreign (passenger) piece of DNA is ligated into a circle with a vector, the essence of cloning has been carried out. However, at least two other competing reactions can occur: reformation of the original circular vector DNA and circularization of the foreign piece. Obviously these are undesirable side reactions, and under most circumstances, conditions are set up to avoid them. Circularization of the vector is favored at low concentrations because the DNA's two complementary ends are always in the same vicinity (they are, after all, on the same molecule). As the concentration of the foreign DNA is increased, reactions between molecules will be more frequent than reactions within the same molecule.

Aside from an increase in concentration, another way of subverting the self-circularization of the vector is by treating its 5' phosphate ends with another enzyme, a phosphatase, that removes a terminal phosphate. Since ligation requires the presence of a 5' phosphate, circularization will not occur.

The phosphatase reaction is illustrated in the adjoining figure. Notice that when the vector DNA is treated with phosphatase, recircularization doesn't occur. Of course, the passenger DNA is not treated with the enzyme. Its 5' phosphates (one on each end) therefore remain intact, and they can be joined with the 3' hydroxyls of the vector. However, note that the resultant recombinant molecule still retains two unligated bonds (single-stranded breaks) at the juncture of the 5' OH's of the vector and the 3' OH's of the introduced DNA. It has been found that despite these so-called **nicks** (a break in the phosphodiester bond of one strand of a double-stranded DNA), the resultant molecules are capable of entering bacteria. And once inside, the nicks can be repaired by enzymes that are normally present within the bacteria.

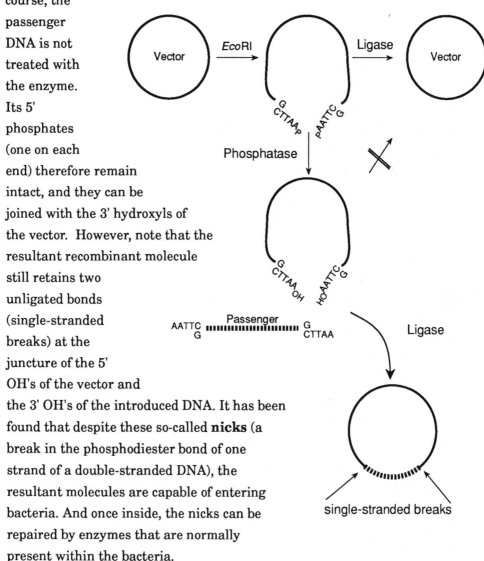

dA/dT and dC/dG joining

A second procedure for joining two DNA fragments requires the addition of complementary homopolymers to their 3' ends. Yet another member of the molecular biologist's toolbox – the enzyme **terminal deoxynucleotide transferase** – is used to carry out the reaction. This enzyme will add nucleotides to the 3' OH ends of a DNA molecule (3' OH sticky ends or blunt ends work best). Normally, DNA synthesis requires a template, but terminal transferase takes whatever nucleotides are around and adds them randomly to the ends of the DNA. If only a single nucleotide is in the reaction mix, a homopolymer will result.

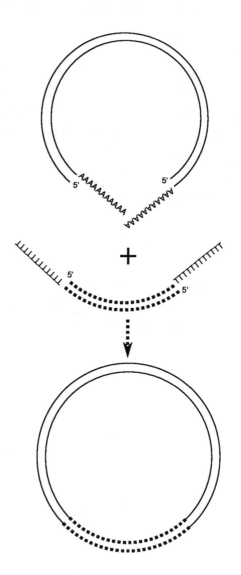

To join two pieces of DNA, poly dT, for example, is added to the 3' ends of one fragment and poly dA to the 3' ends of another. As can be seen in the figure, the two fragments can now readily join because their ends are complementary. Notice that this method, unlike that described in the previous section, cannot produce fragments that circularize spontaneously. In fact, the formation of a circle depends on the introduction of the second fragment.

Blunt-end ligation

Two DNA's can also be joined even if they have *no* regions of single-stranded complementarity, provided that their 5' ends have terminal phosphate groups and their 3' hydroxyl groups are free. Only T4 ligase has this blunt- end ligation activity (a similar enzyme called *E. coli* ligase does not).

Linkers

Another application of the blunt-end ligation activity of T4 ligase is to join synthetic DNA "linkers" to DNA fragments. Linkers are short, symmetrical, self-complementary oligonucleotides that have one or more restriction sites within them. The linker illustrated here is a decamer with an *Eco*RI site within it. One purchases or synthesizes (see below) a specific linker like this one and blunt-end ligates it onto the passenger DNA. Of course, more than one linker molecule may join

$$5' \quad C \ C \ G \ A \ A \ T \ T \ C \ G \ G \quad 3'$$
$$3' \ G \ G \ C \ T \ T \ A \ A \ G \ C \ C \quad 5'$$

on any one end of DNA. There is very little control of the exact number that will be appended by the ligase. However, after the linkers have been added, they may be removed by treatment with a restriction enzyme that recognizes a site within

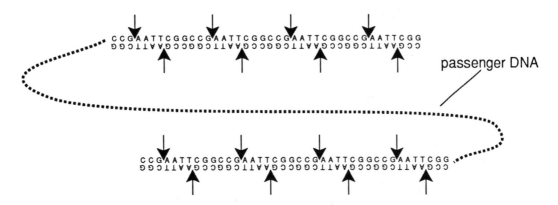

passenger DNA

the linker. This operation removes all but one end of the last linker (see the illustration above) and produces a passenger molecule with the cohesive ends characteristically left by that enzyme. Now the passenger may be inserted into a cloning vehicle that has been cut with the same restriction endonuclease. The advantage of carrying out the reaction in this way is that any passenger molecule can be joined to a vector (not just fragments that happen to have a set of specific restriction sites near its ends) and the fragment can be easily removed with a specific restriction endonuclease.

It has already been noted that to remove excess linkers and to produce the correct ends this procedure requires that the linker-passenger DNA complex be treated with a specific restriction endonuclease. A problem arises when the passenger DNA contains an internal restriction site that is recognized by that enzyme. Instead of being cut just at its ends, the DNA will be cut at this internal site, and more than one fragment will result. In some ligations, both pieces will join the vector in the proper orientation. But in most instances, only one fragment will ligate to the vector, or the two pieces will attach in the wrong orientation. In short, the result will be the cloning of a complex mixture of fragments when only one was wanted.

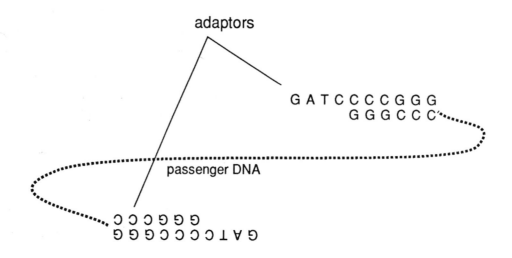

Adaptors

One way around this situation is to use an **adaptor** and to attach it to the passenger. Adaptors are pairs of partially complementary oligonucleotides that upon hybridization form a largely double-stranded molecule with a blunt end and a sticky end. Adapter molecules can be ligated onto the blunt ends of a passenger DNA (as long as they have 5' phosphates), and a specific single-stranded sticky end will result. In the illustration above, *Bam*HI-compatible sticky ends have been added to the blunt-ends of a passenger DNA.

Chemical synthesis

The construction of linkers and adaptors requires that oligonucleotides of defined sequence be synthesized. Today, chemical synthesis of oligonucleotides of up to about 100 or more bases in length has become routine in many molecular biology laboratories or readily accessible central service facilities. The oligonucleotides cannot only be used as linkers and adaptors, but they also serve as primers for *in vitro* mutagenesis and DNA sequencing and as probes for the detection of specific RNA's and DNA's.

The first synthesis of a large oligonucleotide was done by Khorana using the so-called phosphodiester method. Two other chemistries are now widely employed: the phosphotriester and the phosphite-triester methods. In either case, one dramatic technical advance has been to immobilize the growing oligonucleotide chain on a solid support, allowing excess reagents and by-products to be readily washed away. In conjunction with microprocessors and electronically controlled valves, the whole process can be automated. These so-called **gene machines** or oligonucleotide synthesizers can carry out syntheses of polynucleotides up to several hundred bases in length.

Summary

Gel electrophoresis, whether in agarose or acrylamide, enables the genetic engineer to precisely measure the size of DNA fragments of varying size.

Restriction endonucleases are enzymes that cut DNA at defined sites. They leave either sticky or blunt ends.

DNA molecules can be spliced together with the enzyme DNA ligase.

7

Cloning vehicles

Recall that the second step in molecular cloning is to join the passenger DNA with a suitable cloning vehicle. These vehicles (or vectors) must be able to replicate the recombinant DNA formed so that it can be amplified and eventually isolated.

Plasmids

Plasmids are relatively small, double-stranded, closed-circular DNA molecules that exist apart from the chromosomes of their hosts. They are molecular parasites of sorts and are present in a number of different species of bacteria and yeast. Plasmids may carry one or more genes, some of which may confer antibiotic resistance. Some plasmids also bear genes that code for the restriction and modification enzymes that are discussed in the preceding chapter. Some may direct the synthesis of enzymes that aid in the production of bacterial poisons and antibiotics. However, from the viewpoint of the recombinant DNA technologist, the most important property of a plasmid is that it bears a region of DNA – an **origin of replication** – that allows it to multiply within and independently of its host. If foreign sequences are combined with the plasmid, whether they are from sequoias or sea lions, the plasmid doesn't much care. It replicates away, producing many copies of itself and its accompanying passenger.

In a typical cloning experiment, the circular plasmid DNA is linearized by treating it with a restriction endonuclease. Then, as described in the section on ligation in the last chapter, a foreign DNA fragment carrying compatible ends is inserted, thereby producing a circular molecule containing all of the plasmid and its passenger. The next step is transformation – the introduction of a naked recombinant DNA molecule into a single bacterium. Once in, the plasmid replicates as the bacteria themselves grow and reproduce.

Plasmid replication

Many of the plasmids that are commonly utilized in recombinant technology are derivatives of pMB1, a plasmid that was originally isolated from bacteria that had infected a human patient. pMB1 is an example of a **relaxed** replicating plasmid. Relaxed in this case refers not to a state of mind, but to the regulation of the number of copies of the plasmid in the cell. Plasmids under relaxed control are maintained in large numbers – sometimes hundreds of copies – in bacteria. By contrast, there are normally only one or a few copies of **stringently** controlled plasmids in a cell. Furthermore, relaxed replicating plasmids don't require protein synthesis in order to divide. Plasmids under stringent replication do (so does chromosomal replication). Accordingly, in the presence of an inhibitor of protein synthesis (the antibiotic chloramphenicol is often used), relaxed replicating plasmids will accumulate 100-fold until the total plasmid DNA represents up to about half of all cellular DNA. This process is called **amplification,** and it is an often-employed trick for raising the yield of plasmids prior to their isolation.

Desirable properties of plasmids

Plasmids have proven extremely useful as cloning vectors, particularly when they are modified specifically for that purpose. A bacterial plasmid that is designed for cloning should have some or all of the following properties:

• **It should be small.** The aim of most cloning experiments is to isolate the passenger DNA. The vector only serves to carry and amplify the passenger. A small plasmid has the advantage of contributing only a minimal amount of extraneous DNA to the plasmid/passenger construct, thereby making it easier to prepare large amounts of the passenger DNA. In addition, it is easier to get smaller pieces of DNA into bacteria than larger ones – everything else being equal – and the smaller the plasmid is, the less it contributes to total size. Finally, small plasmids are easier to purify than large ones because they are less fragile.

• **Its DNA sequence should be known.** Knowledge of the DNA sequence of a plasmid allows the genetic engineer to manipulate it with the full complement of recombinant DNA techniques. However, if the complete DNA sequence of the plasmid has not been determined, a detailed restriction map should be obtainable. We'll have more to say about restriction mapping in Chapter 10.

• **It should grow to high copy number in the host cell.** In other words, relaxed replication plasmids are most often preferred to those that are under stringent control. In most cases, having many copies makes it easier to purify the plasmid away from the chromosomal DNA, and of course increases its yield. However, there are circumstances when just one or a few copies of a plasmid are desirable. For example, it may be that a particular cloned gene has a detrimental effect on the organism that harbors it. Under those conditions, a few copies of that gene (and hence a few copies of the plasmid) may be less harmful than many.

• **It should contain a selectable marker that allows cells containing the plasmid to be isolated.** Bacterial cells that harbor plasmids don't necessarily look or act unlike those that do not. It is only when the plasmid carries a gene that lends a special trait to the bacteria that the molecular engineer can tell if a bacterium has been transformed. Antibiotic resistance is one such trait or marker, and the genes for ampicillin resistance and tetracycline resistance are common genes carried by frequently used plasmids. Some plasmids are also readily lost from cells, and unless there is a selectable gene contributed by the plasmid and the bacteria are kept under selective conditions, the molecular engineer may end up with a plasmid-less population of cells at the end of an experiment.

• **It should also contain a second selectable gene that is inactivated by insertion of the passenger.** Imagine the following scenario. A scientist is trying to insert a passenger molecule into a plasmid. The plasmid has only a single selectable marker. After ligation, some of the molecules contain the passenger. Others do not. These may be plasmids that have recircularized in the presence of DNA ligase. This mixture of plasmids is then used to transform bacteria. Both recombinant plasmids and recircularized plasmids contain the same marker and cannot readily be told apart genetically. How are colonies carrying a plasmid with a passenger distinguished from those with a plasmid with no passenger? One way is to grow many individual colonies, isolate plasmid DNA from each, run the DNA out on an agarose gel, and identify the recombinant plasmid molecules because they are bigger by virtue of the passenger that they carry. While this procedure works and is very commonly carried out, it is laborious, time consuming, and expensive.

There is a better way. Suppose that a plasmid carries two different antibiotic resistance genes, A and B. What would be the consequences of cloning the passenger into gene B? In many cases, the foreign DNA will disrupt gene B and not allow it to work. On the other hand, if the plasmid simply recircularizes, gene B will be unaffected. With this arrangement it is possible to tell which cells harbor plasmids that carry passengers. Those that contain any plasmid at all will have gene A activity. And those that have a plasmid bearing a passenger, will lack activity from gene B.

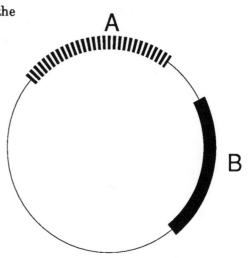

• **There should be a large number of unique restriction sites lying within one of the two selectable markers described above.** Why unique? As described above, most cloning is done by inserting a fragment of DNA that is cut with a specific restriction endonuclease into a vector that is cut with the same enzyme (remember, this ensures that the two have compatible ends). If the vector contains more than one such site, it will be cut into multiple pieces by the restriction enzyme, complicating matters unnecessarily. The presence of many unique sites allows for maximum flexibility and ease in cloning.

Plasmid purification

Plasmids may be purified from bacteria by taking advantage of the difference between the small, circular plasmid molecules and the large, broken (hence linear) pieces of chromosomal DNA. (Chromosomal DNA from *E. coli* is also circular, but during isolation its relatively large size and consequent fragility cause it to break easily.) The most common method for purifying plasmid DNA involves three steps. First, the bacteria are broken open and the is DNA isolated.

Then the DNA is denatured. Finally, the DNA is renatured and centrifuged. Upon denaturation, the two circular single-stranded chains of the plasmid DNA remain entwined and don't separate fully. When conditions are set up so that renaturation can occur, each strand rapidly finds its complement. The chromosomal DNA, on the other hand, breaks readily and therefore consists of noncircular pieces. Under these circumstances, the two strands easily denature and separate. Upon renaturation, they have difficulty finding complete copies of their complements. Often partial renaturation occurs between different sizes of single-stranded fragments and large insoluble aggregates form. These can be simply separated from the small circular plasmid DNA by high-speed centrifugation.

When greater purity is required, plasmids are separated from chromosomal DNA by high-speed centrifugation in density gradients. In this procedure, the plasmids (and contaminants) are placed in very concentrated salt solutions and centrifuged at ten's of thousands of revolutions per minute in 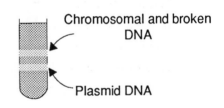 the presence of ethidium bromide. Under these conditions, the plasmid molecules have a different density than linear DNA and are readily separated.

Some popular plasmids

One often used plasmid, pBR322, was pieced together by Francisco Bolivar and colleagues in the 1970s. What makes pBR322 useful is that it contains an

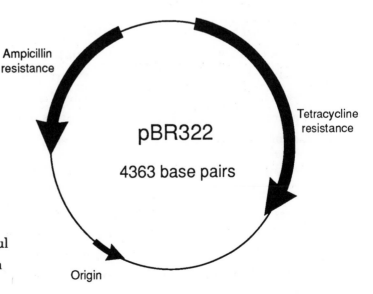

ampicillin resistance gene and a tetracycline resistance gene. In addition it has a relaxed origin of replication and accumulates to high numbers in *E. coli*. Its entire 4363 base-pair sequence has been determined, and 21 common enzymes are available that recognize only a single site within it. (However, only 11 are in either of the two antibiotic resistance genes.)

More recently, a series of small plasmids (about 2.7 kilobase pairs) have been developed that have several properties that have made them very popular with genetic engineers. These pUC (pronounced PUCK) plasmids,

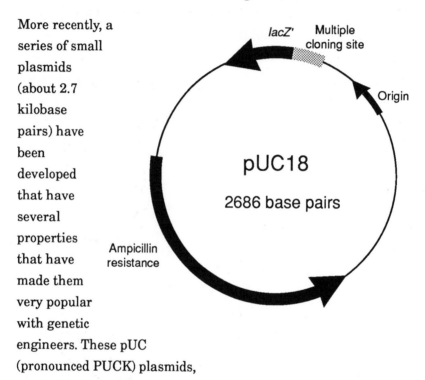

lacZ' Multiple cloning site

Origin

pUC18

2686 base pairs

Ampicillin resistance

exemplified by pUC18 pictured at the right, carry an ampicillin resistance gene and an origin of replication, both from pBR322. They also bear a **multiple cloning site** – a sequence of DNA that carries many restriction sites (13, in the case of pUC18). The multiple cloning site of the pUC plasmids is special because it also codes for a small peptide. This peptide will correct a specific mutation in the chromosomal gene that codes for the enzyme β-galactosidase.

When plasmids containing this sequence are cloned into a specific *E. coli* strain that lacks β-galactosidase activity, they make the peptide and thereby begin to express active enzyme. If, however, one of the restriction endonuclease sites in the multiple cloning site is opened and a foreign gene inserted, the peptide is no longer produced (because the protein coding region is disturbed), and no β-galactosidase activity appears.

Cells that harbor an active β-galactosidase enzyme can be made to turn blue in the presence of certain substrates. Those colonies that have a passenger inserted at the multiple cloning site of the pUC plasmids will lack the enzyme and will be white, while those that have simply recircularized (those that don't contain a passenger) will stain blue. In this way, pUC plasmids containing a foreign insert of DNA can be distinguished from plasmids without a passenger.

Lambdoid phage

Lambda is a temperate bacteriophage with a genome size of about 48.5 kilobase pairs. Its entire DNA sequence is known. In phage particles, the lambda genome exists as a linear, double-stranded molecule with single-stranded, complementary ends. These ends can hybridize with each other (and do so when the DNA is within an infected cell) and are thus termed **cohesive**. They are similar to, but longer than, the sticky ends that were encountered in the last chapter.

Lysis and lysogeny

Bacteriophage lambda can enjoy one of two life styles – that's what's meant by the word **temperate**. It can enter the **lytic** cycle, replicate many times, produce more phage, and destroy its bacterial host. Alternatively, it can take up **lysogenic** growth and integrate into the bacterial chromosome.

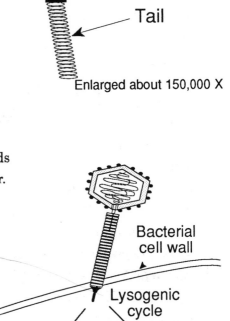

Bacteriophage lambda

Head
DNA
Tail

Enlarged about 150,000 X

Bacterial cell wall

Lysogenic cycle

Lytic cycle

Lambda DNA

Bacterial DNA

The lysogenic state is highly stable, but not permanent. Although the viral DNA may be passed through hundreds of generations of bacteria while integrated into the chromosome, a number of influences can cause "induction" and a return to the lytic cycle. For most purposes of genetic engineering, the tendency of lambda phage to engage in the lysogenic state is a complication that for the most part can be ignored.

Lambda is assayed by plating phage particles on a **lawn** – bacterial colonies spread on plates so that they form a continuous group of cells. If a lambda phage infects one of the cells and enters the lytic mode, it will multiply quickly and lyse that cell, releasing lambda phage. These in turn will infect adjacent cells and also cause lysis. After a while, a round hole – technically, a **plaque** – is seen in the lawn, representing the breakage of many bacterial cells and also marking a single original event – the entrance of a single virus into a single cell.

Shown below is the genetic map of phage lambda (not all the genes are included) and a scale indicating the extent of various regions, expressed as a percentage of the total genome. Notice that related genes are clustered. For example, the genes

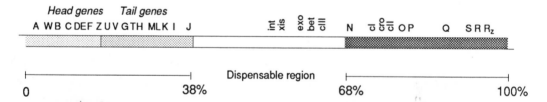

that form the proteins of the head are gathered together toward the extreme left end of the DNA, and the tail-forming genes follow. Similarly, the genes for DNA synthesis (O and P) and those that cause bacterial lysis (S, R, and R_z) are clustered near the right end of the lambda chromosome. In the middle of the chromosome – between the J and int genes – is a large region whose function is unclear but apparently unessential. Also pictured at the ends of the chromosome are the two single-stranded cohesive ends.

Three reasons why bacteriophage lambda is a good cloning vehicle

First, it can accept very large pieces of foreign DNA. About 20kb of DNA can be deleted from its central region (between the J and N genes) and elsewhere, and replaced with an equal quantity of foreign DNA without affecting the ability of the phage to grow lytically. As already discussed in the section on plasmids, a good cloning vehicle should accept foreign DNA and amplify it without regard to the organism from which it comes. Phage lambda fills this bill nicely.

Second, it has been extensively reworked over the years. Genetic engineers have constructed numerous derivatives of lambda that contain only one or two sites for a variety of restriction enzymes.

Finally – its main advantage – bacteriophage lambda is one of the few organisms that can be reconstituted in a test tube. By simply mixing phage DNA with a mixture of phage proteins, an infective viral particle with the DNA inside the phage head can be produced. This process is called *in vitro* **packaging** (or just plain **packaging**) and it normally occurs when long tandem arrays of phage chromosomes (they are produced in this form when they replicate in the bacterial cell) are cleaved into monomers by certain phage proteins. The DNA is cut at a site on the chromosome called *cos*, and this operation produces the cohesive ends that were mentioned above. Ordinarily the DNA is cut into 48.5kb fragments that are incorporated into the head of newly formed phage particles.

However – and this is very important – there is a strict size requirement for the piece of DNA that goes into the head. That is, if the distance between successive *cos* sites is either too long or too short (longer than about 105% or shorter than about 78% of a wild-type phage's DNA), the resultant phage will have markedly decreased viability. For example, if a piece of phage DNA of 36kb (about 75% of 48.5kb) is packaged, it will fail to yield significant numbers of active phage. The same thing will happen if the distance between two successive *cos* sites is 53.4kb, approximately 110% greater than the normal size of a lambda chromosome.

The molecular engineer can take advantage of this situation to distinguish among recombinant molecules that carry foreign DNA and those that do not. The essential

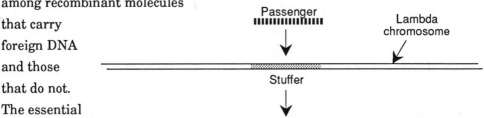

Passenger

Lambda chromosome

Stuffer

idea is to begin with a vector that carries a piece of useless DNA (a so-called **stuffer** fragment) that can be excised with the aid of one or more restriction endonucleases. The stuffer fragment is specially designed so that when it is removed, the spacing between the *cos* sites will be too short for successful packaging. If a passenger is not substituted in place of the stuffer, no infective phage particles will be produced.

Phage that have a stuffer fragment are called **substitution vectors** because they are designed to have a piece removed and substituted with something else. A second class of phage lambda cloning vehicles are called **insertion vectors**. They behave similarly to the plasmid vectors that we have already considered – foreign DNA is simply inserted into an opened site and no stuffer is thrown away.

Large numbers of different lambda strains have been created that allow efficient cloning of a variety of foreign DNA's. The various strains are designed to have differing amounts of DNA removed, and they contain a variety of restriction enzyme sites for cloning. In addition, some lambda strains have a stuffer fragment that carries the β-galactosidase gene. When it is removed or when foreign DNA is cloned within the gene, β-galactosidase activity may be abolished. The accompanying loss of activity may be used to select recombinant clones.

Why lambda?

Why use bacteriophage lambda as a vector rather than plasmids? Large pieces of DNA (up to about 20 kilobase pairs) can be easily cloned in bacteriophage lambda substitution vectors. Plasmid vectors are less useful for cloning big passengers.

But why clone large pieces of DNA in the first place? One obvious reason is that some genes are very big and it is advantageous to have them all in one piece. Interestingly, often it is not the protein-coding portion of these large genes that contributes to their size. Rather, the regulatory regions of several genes are dispersed tens of thousands of base pairs away from the structural gene. If the goal of a cloning experiment is to inject a particular gene into a host and if that gene must act properly there, it is necessary that the gene carry its regulatory region with it. Some genes are large because they carry **intervening sequences** (or **introns**) in them. The topic of intervening sequences is discussed further in Chapter 8.

Another reason for cloning in lambda is the efficiency it offers in DNA transformation. In general, plasmids are much less efficiently moved into bacterial cells than bacteriophage, which after all were designed by Nature for their ability to project their chromosome into their hosts.

Cosmids

Cosmids, as the name implies, are plasmids that carry one or more *cos* sites from bacteriophage lambda. Consequently, they share some properties with both plasmids and lambda phage. The presence of the *cos* sequence allows cosmid DNA to be packaged. (Recall that the phage lambda packaging system will successfully put any DNA into a viable phage as long as it carries successive *cos* sites separated by about 50 kilobase pairs of DNA.) Once it gets into a bacterium, the cosmid behaves like a plasmid – it doesn't have any other lambda-derived sequences. It multiplies under the control of a plasmid origin of replication, and it may carry selectable genes just like a plasmid. It is also subject to amplification just like plasmids.

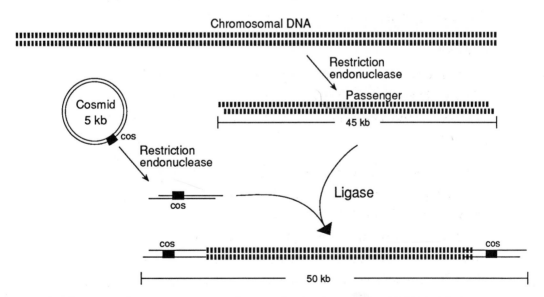

As shown in the accompanying figure, cloning passengers in cosmids is conceptually quite simple. Two cosmids are ligated to the passenger – one on each end – to create a molecule with two *cos* sites separated by the passenger. This molecule is then placed in a packaging extract and from there inserted into bacteria just as if it were lambda DNA.

The major reason to turn to cosmids for cloning rather than plasmids or lambda phage is that cosmids can be used to efficiently clone large pieces of DNA. This ability arises from the observation that the lambda packaging system can put nearly 50 kilobase pairs of DNA into phage heads. Coupled with the fact that cosmid cloning vectors can be made quite small, the result is the ability to clone about 45 kilobase pairs of passenger DNA (the difference between cosmid size and phage lambda head capacity). As we'll see in Chapter 8, cosmid cloning is a handy way to make libraries of sequences from organisms with large genomes.

Filamentous phage

Filamentous phage (M13 is a prime example) produce infective particles containing only a single strand of DNA of about 6.4 kilobases. Upon infection, the single-stranded chromosome of these phage enters *E. coli* and is converted to a double-stranded, circular form. It then multiplies rapidly until it reaches a steady-state number of 50 to 200 molecules per cell. At this stage, M13 DNA is behaving like a plasmid, and in its double-stranded form it may be purified from cells and manipulated as such.

Bacterial cell membrane

Mature phage protein coat

INSIDE   OUTSIDE

But 15 minutes after infection, M13 begins synthesizing single-stranded circular DNA: molecules that are destined to go into viral particles. This DNA gets covered with proteins while in the bacterium, but formation of the mature phage particle occurs only when it reaches the bacterial cell membrane. Here the phage chromosome becomes encoated with a new set of proteins while it is being secreted to the outside. This packaging, unlike that of bacteriophage lambda, is not dependent on size. Also unlike lambda, filamentous phage manage to exit without lysing their hosts (although they do slow down the rate of growth of the bacteria and thereby produce plaques). Because the phage particles are secreted (and hence easy to purify by simply centrifuging the bacteria out of the culture medium) and can accommodate variable lengths of DNA, M13 is an ideal vehicle for cloning single-stranded DNA's.

What are single-stranded DNA's good for? For one thing, they are useful for DNA sequencing using the Sanger dideoxy method (see Chapter 10). They have also been used to fish out specific mRNA's from a mixed population (see Chapter 10) and for the production of single-stranded probes (Probes are covered in Chapter 9). In addition, methods for *in vitro* mutagenesis using M13-derived DNA have been developed and these have increased the popularity of this vector.

Yeast vectors

Baker's yeast (technically *Saccharomyces ceriviciae*) is playing a growing role in molecular genetics in both basic research and industrial applications. One reason is that it has many of the characteristics of so-called higher organisms, despite the fact that yeast cells are small and the organism isn't multicellular. Yeast have, for example, true nuclei and therefore belong to the group of organisms called **eukaryotes** (as contrasted with bacteria which are **prokaryotes**). Like mammals, they have multiple linear chromosomes. They also undergo mitosis and meiosis. And they even contain mitochondria and other familiar organelles. But most interesting (and somewhat unexpected) has been the discovery that yeast have many genes that are homologous to those of humans and other multicellular organisms. In some cases, a corresponding yeast gene can even substitute for a human one and vice versa. Because of this homology, biologists are learning about some human genes by studying the behavior of the corresponding yeast gene in yeast or, in some cases, by putting the corresponding yeast gene into human cells.

All yeast vectors are plasmids. Most are derived from pBR322. They come in a number of different varieties termed YIp, YEp, YRp, and YCp, which stand respectively for yeast **integrating, episomal, replicating**, and **centromere** plasmids. Because it's easier to grow and purify plasmids from *E. coli*, all are capable of growing in both bacteria and yeast (hence the term **shuttle vectors**). To this end, they all have a bacterial plasmid origin of replication, which allows them to multiply in bacteria. They also bear one or more bacterial genes that can serve as selectable markers. In addition, they usually have one or more yeast genes that can be recognized when the plasmid is introduced into *Saccharomyces*. (The *URA3* gene is used as an example in the figures. When introduced into *ura3* mutants, it confers the ability to grow on media lacking uracil.) Finally, as with most useful vectors, they carry a number of unique restriction sites.

YIp's, YEp's, YRp's, and YCp's

The four types of yeast plasmids differ primarily in how they replicate in yeast. Recall that all the plasmids carry a bacterial origin, but it isn't recognized by the yeast replication machinery. If a plasmid has no yeast origin (YIp plasmids), it cannot replicate after it gets into the yeast cell. When the yeast cells multiply, it is diluted out and effectively lost. The only way that it can be stably maintained in the cells is by integration, generally by recombining with a sequence on one of the yeast chromosomes. Since

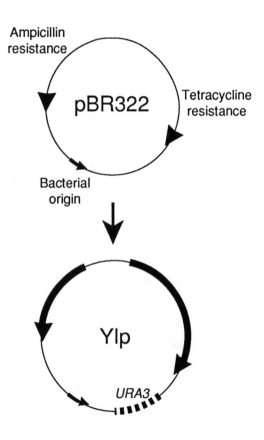

recombination is a relatively rare event, YIp plasmids have low efficiencies of transformation. When they do integrate, usually by recombination, only one or a few copies participate. But once integrated, they are stably carried from one generation to another.

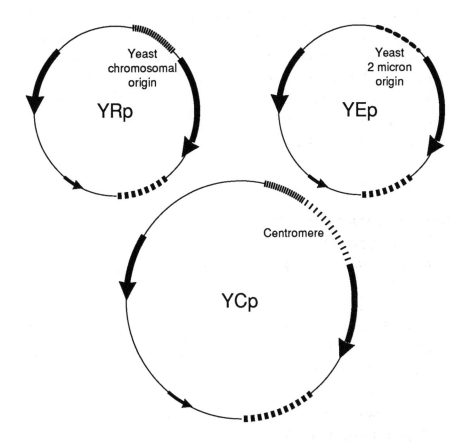

By contrast, the YEp plasmids carry an origin of replication of a naturally occurring plasmid of yeast (a plasmid called the **2 micron circle** that is present in multiple copies in many yeast strains). YEp plasmids replicate many times per cell generation and accumulate to high numbers (25 to 100 copies per cell). Because of their ability to replicate readily when they enter a cell, they transform yeast at high efficiencies. And because of their high copy numbers, genes carried by YEp plasmids may be expressed in large quantities. Finally, the presence of so many copies of the plasmid in a cell increases the chances that they will be passed on during mitosis. Hence they are stably maintained even in the absence of selection.

Somewhere in between YIp's and YEp's are the YRp plasmids, which contain an origin of replication normally found on the chromosomes of yeast. When introduced into cells, they multiply to moderate copy numbers (less than YEp plasmids) and sometimes integrate into chromosomes. If they do not integrate, they tend to be lost during division of the yeast cells.

Finally, there are the YCp plasmids, which contain a chromosomal origin and a yeast centromere sequence (a DNA sequence that ensures that chromosomes move to opposite poles during mitosis and meiosis). They transform yeast at moderate rates but only a few copies enter and are maintained in each cell. The centromere tends to allow YCp plasmids to segregate properly during mitosis, and they are not as readily lost from cells as are YRp's. However, they are still much less stable then ordinary chromosomes (or genes that are integrated onto chromosomes).

The general strategy used with all these yeast vectors is to isolate large quantities of the plasmid from bacteria, insert the desired genes by the methods described in Chapter 6, grow the passenger/vector recombinant in bacteria, and then introduce the modified plasmids into yeast by transformation. Once in yeast, they may be studied exhaustively, using the many powerful tools of yeast genetics and molecular biology. The passengers can also be made to express proteins, a topic that we will come back to in Chapter 11.

YAC's

Bacteriophage lambda and cosmids can bear large fragments of DNA. Perhaps the ultimate vehicle for this purpose is one recently developed for use in yeast. These plasmids – called **YAC (yeast artificial chromosomes) vectors** – are capable of carrying segments of foreign DNA up to an order of magnitude larger in size than a cosmid (500kb). As indicated in the figure on the next page, they are similar to YCp plasmids, but considerably more complicated. As such, they have three selectable yeast genes, a yeast chromosomal origin of replication, and a yeast centromere in addition to the usual bacterial plasmid origin and markers. Two segments of DNA that mimic the ends (**telomeres**) of yeast chromosomes are also present.

To clone a gene into a YAC vector (and thereby produce a YAC), the plasmid is cut with two restriction enzymes, one that removes a stufferlike fragment that separates the two telomere sequences and another that cleaves within one selectable marker (designated gene 1 in the figure). This maneuver breaks the vector into two pieces, each of which bears a telomere on one end and one of the two remaining selectable genes (genes 2 and 3 in the figure). When these two pieces are ligated to the ends of a large piece of passenger DNA and when this linear DNA is transformed into yeast, it produces something that looks and behaves exactly like a yeast chromosome.

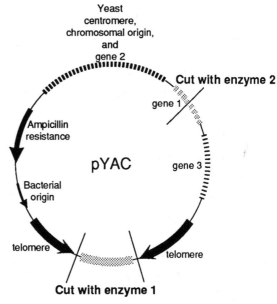

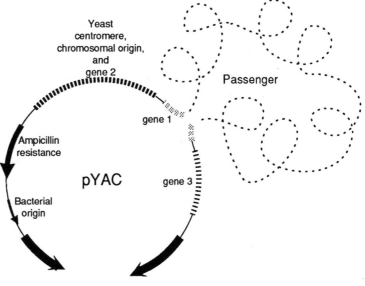

YAC's may allow scientists to study genes that are unclonable because of their large size in any of the vectors previously described. They may also allow groups of genes to be linked together creating the possibility that whole metabolic pathways may be introduced into yeast.

Summary

The second step in molecular cloning is to join the passenger DNA with a cloning vector.

Plasmids are closed-circular, extrachromosomal DNA's that are frequently used for cloning relatively small pieces of DNA.

Bacteriophage lambda DNA carries a dispensable region that may be substituted by a foreign passenger. Lambda is capable of being reconstituted in a test tube.

Cosmids share some characteristics of plasmids and phage lambda. They are capable of carrying large fragments of passenger DNA.

The filamentous phage are useful for cloning DNA's that are required in single-stranded form.

Yeast plasmids may be used when expression of a foreign gene in a eukaryotic cell is needed. Yeast artificial chromosomes (YAC's) are useful for cloning extremely large pieces of DNA.

8

cDNA and genomic libraries

A brief review of what's been covered so far... We've seen that DNA can be cut into discrete pieces with restriction endonucleases. These passenger molecules can be linked via DNA ligase to one of a variety of vectors and propagated in bacteria or yeast. The result is a collection of molecular clones – recombinant DNA's – the ultimate object of the cloning procedure.

However, the process is not yet complete. If a genetic engineer begins with a mixture of passenger molecules, he or she will end up with a mixture of recombinant DNA's. While identical copies of a single hybrid molecule will be present in any individual bacterial colony or viral plaque, there may be millions of such colonies if there were millions of different passenger molecules at the start. If a specific gene is sought, how can it be found in this jumble?

This critical question is addressed in the next chapter. For now, note that the initial cloning of most genes is indeed done from complex mixtures of cloned passenger fragments. These mixtures are called **libraries** or **clone banks** and they may carry either cloned cDNA's or genomic passengers.

cDNA's

A cDNA is a double-stranded copy of an RNA. cDNA libraries are collections of cloned cDNA's that arise from a mixture of RNA's. As discussed later in this chapter, it is relatively easy to build such collections of cDNA's in a few enzymatic steps in a test tube. But why do so? Why clone DNA copies of RNA when the real thing can be obtained directly from the genome?

There are several compelling reasons. For one thing, there is the possibility that a collection of cDNA's may carry a particular gene of interest in a higher concentration than the same gene in the genome, thereby making it easier to identify and purify that gene. For another, cDNA's may have some properties that genomic DNA lacks. Each of these points is considered in turn.

Ease of purification

Some tissues accumulate massive amounts of one particular mRNA species. Because of its relative abundance, it may be easier to purify this mRNA and make a cDNA out of it than to isolate the corresponding gene from a genomic digest.

Ease of isolation

Eukaryotic mRNA has another property that simplifies its isolation. Nearly all eukaryotic mRNA's carry a string of 50 to 200 A residues at their 3' ends (they are said to be **polyadenylated**). These A's can serve as molecular handles that can be grabbed by a complementary string of U's or T's attached to a solid support. In this way mRNA molecules can be readily separated from the more abundant nonpolyadenylated ribosomal and tRNA's.

The DNA doesn't have to be cut from larger pieces

Sometimes a genetic engineer will find it difficult to cut out a complete copy of a whole gene from the genome. For some purposes – the synthesis of a complete protein is one – the cloning of only a portion of a gene is unsatisfactory. cDNA cloning can alleviate this problem. Since mRNA's come in definite sizes, they do not have to be excised from the rest of the genome with restriction endonucleases.

Introns are already removed

In higher eukaryotes, many genes contain sequences that are removed from the initial transcript. As a result, a mRNA may be missing some sequences that were present when the gene was transcribed, and therefore a cDNA copy of it may not represent an accurate version of the corresponding DNA. These missing sequences are called **intervening sequences** or **introns** (the parts of the gene that remain in the mRNA are called **exons** – for expressed sequences). Some genes may have zero, one or two introns; others may have scores. In fact, some genes may consist mostly of introns. The introns may be so large and so numerous that they make it difficult to clone the entire gene in one piece in a cosmid or lambda vector. Because introns are removed from mRNA, a cDNA will certainly be smaller than the genomic copy and may prove more convenient to work with.

The removal of introns is also useful if the goal of an experiment is to translate the product of a gene in bacteria. Introns interfere with translation because they are not properly removed from transcripts by the bacterial cell: Bacteria appear to lack the necessary apparatus. Since cDNA's represent copies of processed RNA, the introns have already been removed, and translation will proceed properly in bacteria.

The process of cDNA cloning

The synthesis of a double-stranded cDNA is a multistep enzymatic synthetic process. Many different procedures have been developed for carrying out the enzymatic steps involved and for cloning the resultant DNA. One older – some might say classic – procedure is outlined below.

First, the genetic engineer isolates the mRNA transcript of the gene of interest. For purposes of illustra-
tion, I will

mRNA 5' ■■■■■■■■■■■■■■■■■■■■■■■■■■■■■■ AAAAAAAAAA... 3'

assume that the mRNA is from a eukaryote and is polyadenylated. The row of A's drawn at the end of the mRNA represents a polyadenylated 3' end, which typically will be longer than that shown in the figure.

Second, a single strand of DNA complementary to the RNA is synthesized using the enzyme reverse transcriptase. The result is a cDNA, albeit single-stranded. Reverse transcriptase is a DNA polymerase, and it has the general properties characteristic of these enzymes (see Chapter 4). For

mRNA 5' ■■■■■■■■■■■■■■■■■■■■■■■■■■■■■ AAAAAAAAAA... 3'
cDNA ▬▬▬▬▬▬▬▬▬▬▬▬▬▬▬▬▬▬▬▬▬▬▬▬▬ TTTTTTTTTTT...
3' 5'

example, it requires a primer, and it adds nucleotides to the primer's 3' end.

However, the enzyme differs from the DNA polymerase previously discussed in
that it utilizes RNA, instead of single-stranded DNA, as a template. As shown in
the figure, the primer used here is poly T, which hybridizes to the poly A tail of
the mRNA.

Third, the mRNA that remains is removed by degradation with alkali (the phosphodiester bonds in RNA are sensitive to highly alkaline solutions). Notice that
after removal of
the RNA, the
cDNA is shown
forming a hairpin
loop at its end, a reaction that typically seems to occur at the completion of this
step.

Fourth, a complementary second strand of DNA is synthesized via *E. coli* DNA
polymerase I.
This second
strand is
primed by
the hairpin
loop formed as a result of the last step.

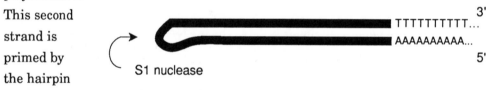

Fifth, the hairpin
loop is removed
using an enzyme
that is cuts DNA
specifically at single-
stranded regions: S1 nuclease.

Finally, the resultant double-stranded cDNA must be inserted into a suitable
vector. Cloning is done usually by adding either a homopolymer tail to the vector
and cDNA or by adding linkers. These steps are described in Chapter 6.

Changes to the procedure

Many variants of the cDNA cloning procedure have been developed that increase
its efficiency and eliminate many of the problems often encountered. A difficulty

that comes up frequently occurs at the first step. Quite often the synthesis of the DNA copy fails before it proceeds all the way to the 5' end of the template. One reason that this occurs may be because some commercial preparations of reverse transcriptase are contaminated with varying amounts of RNase – an enzyme that degrades RNA's and is notoriously difficult to get rid of. If care is taken to purify the reverse transcriptase enzyme or if good inhibitors of RNase are included in the first two steps of the procedure, longer cDNA copies tend to be formed.

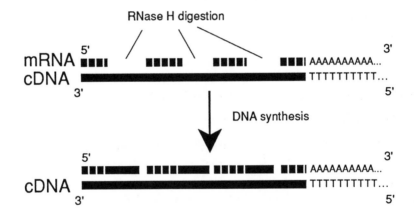

Another serious impediment to getting a full-size cDNA is the treatment with S1 nuclease, a step that is required for removal of the hairpin loop. Since S1 nuclease is a degradative enzyme, it removes deoxynucleotides and, by definition, must therefore shorten the resultant product. To bypass this step, genetic engineers have turned to replacement synthesis for production of the second DNA strand rather than extending the hairpin loop as previously described. In this procedure, the DNA-RNA hybrid is first treated with RNase H, an enzyme that creates gaps in RNA in a DNA-RNA hybrid. Each of these can serve as a primer for DNA synthesis, as shown in the figure. Treatment of these short segments of DNA with DNA ligase joins the pieces and produces a complete second strand.

Genomic libraries

Genomic libraries consist of a series of random DNA sequences from an organism of interest. Often these libraries contain a collection of bacteriophage lambda clones because of the phage's ability to incorporate large pieces of passenger DNA. Cosmids, which are capable of cloning even larger pieces of DNA, are also used frequently.

A genomic library is made by cutting the target DNA into pieces either by shearing or with restriction enzymes. The pieces are then introduced *en masse* into an appropriate vehicle, for example, a bacteriophage lambda substitution vector. After packaging, the recombinant phage are mixed with an excess of bacteria such that only a single phage infects any given bacterium. Under these conditions, each plaque that forms represents a clone of bacteriophage; each phage particle in that plaque containing an identical copy of foreign DNA integrated into the lambda genome.

Producing a complete library

Let's consider some of the problems involved in constructing a genomic library. For ease of calculation, it is assumed that the library is being made from a hypothetical organism that has a genome size of one million base pairs, all on one chromosome.

The genetic engineer begins by extracting DNA from a large number of the organism's cells. To ensure that the library is complete, every nucleotide sequence must be represented at least once in the library. In this case, it should be immediately obvious that the entire genome could fit into 50 lambda clones, each carrying a 20,000 base-pair insert (50 times 20,000 = 1 million). But how can the DNA be broken precisely into 20,000 base-pair pieces? Shearing the DNA by mechanical means might be one way, but the DNA breaks randomly, and pieces will invariably overlap. Consequently any collection of 50 clones of the requisite size will invariably contain gaps and duplicated regions. On the other hand, if the DNA is cut with a restriction endonuclease, the sites that it recognizes will almost certainly not be precisely 20,000 base pairs apart. There will be some sites that are very close together and others that are quite far from each other. That means that pieces that are too big or too small to be cloned in lambda phage will be produced. And again some sequences will be missed.

The solution to this problem is to make a large collection of randomly cut pieces. If the collection is big enough, even if there is considerable overlap among some pieces, the chances are good that at least one clone in the collection will carry a given part of the genome. These randomly cut pieces can be produced either by shearing the DNA or, more commonly, by digesting it with a restriction endonuclease that cuts very frequently. Enzymes that recognize a four base pair site (see Chapter 6) are ideal for this use. The trick is to do an incomplete digestion. By treating the DNA with just a little enzyme or by digesting for only a brief time, the DNA will be cut rather infrequently at a small sampling of a large number of potential sites, and pieces of 20,000 base pairs can be isolated and used to construct the library.

How are these large pieces of DNA purified? A traditional approach had been sucrose gradient centrifugation. Here a mixture of pieces of different size are subjected to ultracentrifugation in tubes containing increasing concentrations of sucrose. The DNA is layered on the gradient and centrifuged so that the larger pieces move further down the tube than the smaller ones. In some instances, sucrose gradient centrifugation has been replaced by agarose electrophoresis.

Getting the recombinant DNA into cells

After making the recombinant molecules, they must be placed into appropriate hosts – the last step in the cloning procedure. Recombined plasmid DNA is generally introduced into bacteria by simply adding DNA to cells that have been treated with calcium chloride. This process is called **transformation**, a term that is generally used to describe the process of getting naked DNA into cells of all kinds. Calcium chloride induced transformation is not a high-efficiency process. Typically with *E. coli*, the technique produces about a million or so transformants per microgram of plasmid DNA.

Cosmid and recombined lambda DNA can be injected into bacteria by bacteriophage lambda particles that have packaged the DNA. Because the phage is designed by Nature for inserting DNA into cells, it is much more efficient at getting DNA in than transformation. Up to 100 million plaques can be produced per microgram of phage DNA.

Summary

Collections of cloned fragments of DNA are called libraries or banks They may consist of either cDNA or genomic clones.

cDNA clone libraries are derived from collections of mRNA that have been converted into DNA by the enzyme reverse transcriptase.

Genomic clones are commonly constructed by partial restriction digestion of nuclear DNA with an enzyme that recognizes many sites. Bacteriophage lambda and cosmids are two vectors that are often employed when generating a genomic library.

9

Selection for the right fragment

As described in the last chapter, large libraries of cloned DNA's can be produced either from cDNA's or from the genome. Many different sequences may be contained with these collections, but if the library is large enough, there will be a high probability that any specific DNA sequence will be included. This chapter discusses the vexing question of how to sort through the collection – how a particular plaque or colony of cells that contains a fragment of interest can be located and distinguished from surrounding ones.

As we'll see, there are many techniques available for this purpose. One powerful and often employed method is **nucleic acid hybridization**. But to detect a specific fragment of DNA by hybridization, a labelled probe is required. The following section explores how probes of this sort are fashioned.

Labelling probes

A hybridization probe is a segment of nucleic acid that is complementary to all or part of the DNA that is being pursued. In most cases the sought-after DNA is fixed onto some solid support (like a nitrocellulose filter), and the probe is added in a solution that bathes the filter. Temperature and salt concentrations are arranged so that hybridization (complementary base pairing) can occur between the probe and its complement. Subsequently, unhybridized probe is washed away, and only hybridized probe remains bound to the filter.

To detect the probe it must be labelled; that is, it must be marked with some signalling device that allows it to be easily detected. Traditionally, the label that is used most frequently is radioactive ^{32}P. Its radioactivity (and hence the presence of the probe) is detected by autoradiography: the blackening of photographic film by particles given off by the decaying phosphorus atoms.

Recently, nonradioactive labels like **biotin** have begun to replace radioactive ones. Biotin is a small molecule – a vitamin – whose major asset for labelling purposes is that it forms a very tight complex with a protein called **avidin**. (Avidin is found in hen eggs in high concentration. If someone consumes too many raw egg whites, the avidin can complex with biotin, remove it, and cause a vitamin deficiency.)

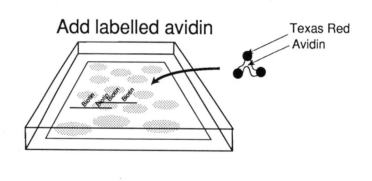

Once incorporated into nucleic acids, biotinylated probes are localized indirectly. The sample to be probed is bathed with a solution containing avidin. The avidin, in turn, binds to any biotin that may be present. The avidin is usually purchased (or modified in the laboratory) so that it carries some signalling molecule – a fluorescent dye or enzyme, for

example – attached to it. (Texas Red, in the example shown at the right, is a commonly used fluorescent dye). What is ultimately detected is this label. The basic idea, however, is that a biotinylated probe can be monitored with great sensitivity.

Nick translation

Whether a radioactive or enzymatic method of detection is used, the probe must be labelled by introducing either radioactive or biotinylated nucleotides into its sequence. One convenient way of doing this is by a process called **nick translation**.

Nick translation, not to be confused with protein translation, means the movement of a nick – a break in the phosphodiester backbone of one strand of double-stranded DNA – from one position to another. The movement is catalyzed by the *E. coli* enzyme **DNA polymerase I**, which recognizes the nick and begins synthesis of a complementary strand when it encounters it. If

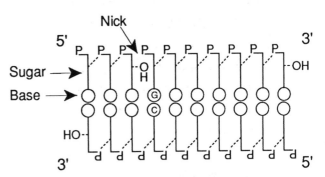

Step 1
DNA polymerase I removes the nucleotide at the nick by breaking the bond to the next nucleotide.

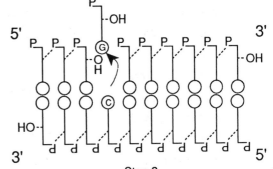

Step 2
DNA polymerase I replaces the missing nucleotide with a new one, leaving a nick at the next nucleotide.

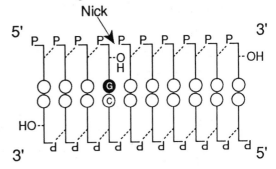

radioactive or biotinylated nucleotides are present during synthesis, the newly
synthesized DNA will become labelled.

Overall, the labelling proceeds as follows: First, DNase is used to introduce
single-stranded gaps in the DNA that is to be labelled. Then DNA polymerase is
put to work. The enzyme removes the nucleotide at the nick by breaking the
bond between it and the adjoining nucleotide. Then it inserts a new nucleotide,
forming a bond between its phosphate and the 3' OH group. This leaves a new
nick one nucleotide over. The result is the apparent movement of the nick in a 3'
direction. (See the diagram on the previous page.)

End labelling

Nick translation is a
powerful, general method
for labelling probes
throughout their length.
Sometimes, however, it
is desirable to

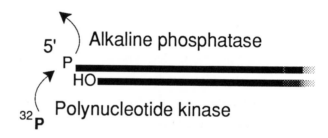

specifically label only the ends of a DNA fragment. For example, having a label
on the ends of a DNA molecule can simplify restriction mapping (Chapter 10). A
number of methods are available for accomplishing end labelling, and the choice
of which one to use depends on the type of overhang: whether blunt or sticky
ends, and whether the sticky ends have a 5' phosphate overhang or a 3' OH.

When faced with a sticky end with a 5' phosphate overhang or with a blunt-
ended fragment, the phosphate may be removed with the enzyme **alkaline
phosphatase** – a part of the genetic engineering tool kit that we've already
encountered. It then may be replaced with radioactive phosphate using the
enzyme **polynucleotide kinase** from the bacteriophage T4 to catalyze the
reaction.

Overhanging 3' OH ends have recessed 5' phosphate groups. While these
phosphates can be removed by alkaline phosphatase, T4 polynucleotide kinase
does not label the resulting ends as well as when they are exposed. To achieve
higher efficiency labelling, some genetic engineers use the previously
encountered enzyme deoxynucleotide terminal transferase to add a string of

radioactive ribonucleotides (not deoxyribonucleotides) to the overhanging 3' OH ends. Because bonds between ribonucleotides can be cleaved with a base, all but the last can be removed by treatment with alkali.

Using the labelled probe

With the labelled probe in hand, it is possible to use colony or plaque hybridization – two techniques that are described later in the chapter – to distinguish wanted clones from worthless ones. But, readers who have been carefully following the train of argument advanced so far may have a question. If the goal is to find a particular piece of DNA by hybridization, doesn't that require a probe with a complementary sequence? And if that's so and the sequence has already been obtained, why clone it? And if it hasn't, how is it derived? These questions are perceptive ones. The fact is that obtaining an appropriate probe is often the most difficult part of the cloning procedure.

Homologous probes

One approach is to use a homologous probe from a closely related organism. In this case, the molecular engineer contacts a colleague who has already cloned a homologous gene and asks for a plasmid or lambda clone carrying the appropriate passenger. The passenger is then employed as a probe to identify a desired gene. For example, it has been possible to detect chicken actin gene-containing clones in a library made from chicken genomic DNA by making use of a probe made from an isolated mammalian actin gene. This protocol is called **cloning by phoning**.

Homologous probes work best when they are very similar to the sequence being sought. However, by lowering the **stringency** of hybridization (allowing some mismatches in base pairing between the probe and its target), it is possible to use more distantly related genes as probes.

cDNA probes

A second approach is to use DNA from a cDNA clone as a probe. As mentioned above, in some cases it is possible to find a tissue or stage of development in which a particular mRNA is present in high concentrations. A labelled cDNA made from a preparation of RNA that is rich in a specific mRNA can work particularly well as a probe for finding a gene in a genomic library. In some cases, cDNA cloning may not even be necessary. If the concentration of a desired messenger RNA is high enough in a particular tissue or at a specific stage of development, it may be possible to make labelled cDNA's from that tissue or to label the RNA directly and use either of these to probe a genomic library.

Back translation

Proteins were purified and their amino acid sequences were determined long before the recombinant DNA revolution. In recent years, new advances in technology have made it possible to determine the sequence of proteins that can only be obtained in minute amounts. Modern machines carry out the process automatically and rapidly.

Armed with the sequence of a protein whose gene is being sought, **back translation** can be used to formulate the sequence of an appropriate probe. Back translation takes advantage of the fact that the triplet genetic code has been deciphered (see Chapter 3) and that the same code is shared by most organisms. One simply substitutes the corresponding three nucleotides for each successive amino acid in the protein sequence.

However, there are some difficulties with this approach. If every amino acid were coded by a single codon, it would be a simple matter to use the genetic code to deduce the precise sequence of any gene from the amino acid sequence of the corresponding protein. But the problem is that most amino acids are represented by more than one set of triplets. Because of this degeneracy, investigators who want to make probes via back translation try to choose parts of proteins that are rich in those amino acids that are directed by a single code word, such as methionine and tryptophan. They try to avoid sections of the protein that are rich in leucine and arginine (for example), which are coded for by 6 different triplets. If they succeed in finding such a region, the deduced nucleotide sequence is synthesized with a gene machine, and the resultant oligonucleotide is labelled and used as a probe.

Colony and plaque hybridization

These procedures are used routinely to identify desired genes in a library once a nucleic acid probe is available. Colony hybridization is used for locating particular plasmids in bacterial colonies.

First, the bacteria are plated onto regular bacterial medium in a Petri dish. After incubation overnight, the bacteria form colonies. Then a nitrocellulose filter is placed on the colonies. A fraction of the bacteria from each colony are transferred to the filter in this process. The original agar plate is then set aside, and subsequent treatments are done to the bacteria on the filter.

The next step in the procedure is to break open the bacteria and release their DNA onto the filter by treatment with base. NaOH This lysis procedure is set up such that the DNA from each colony sticks to the filters at precisely the position where it was liberated. If a particular colony carries a specific passenger DNA in a plasmid, it can be detected by hybridization

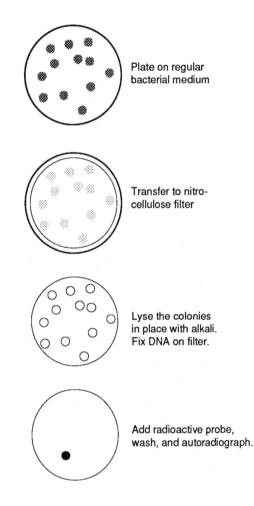

Colony hybridization

Plate on regular bacterial medium

Transfer to nitro-cellulose filter

Lyse the colonies in place with alkali. Fix DNA on filter.

Add radioactive probe, wash, and autoradiograph.

to a specific spot on the filter with a labelled probe. Radioactive probes can be detected by autoradiography. Biotinylated probes can be detected as described above. In either case, colonies are recovered by going back to the original plate, aligning the spot with a colony, and replating that colony.

A similar procedure for detecting specific sequences in recombinant lambda plaques has also been worked out. Again a nitrocellulose filter is used. It is carefully placed over a lawn of bacteria containing lambda plaques, and then, after the DNA is denatured, the filter is treated with a hybridization probe.

Isolation of genes by functional assay

What happens if you can't clone by phone or perhaps it proves difficult to obtain a decent cDNA probe? What can you do if you don't even know the amino acid sequence of the protein encoded by the gene you are looking for?

Complementation

Several different yeast genes have been cloned in bacteria because of their ability to complement a bacterial function. For example, some *E. coli* strains are defective in some of the genes required for the synthesis of the amino acid histidine. In a few cases, the corresponding yeast genes have been cloned because they were capable of expressing a similar enzyme and making up for the defect. The cloning was done by generating a yeast genomic library in a bacterial strain incapable of growing on medium lacking histidine. By simply plating these bacteria onto culture medium lacking histidine, it was possible to select for those colonies that harbored a plasmid with a yeast gene capable of correcting the defect. However, this method isn't extensively used because there are so few eukaryotic genes that complement similar ones in bacteria.

Immunochemical methods

Another much more general and powerful procedure uses the products of the vertebrate immune system. If a nucleic acid probe isn't available, it is possible to recognize a specific gene in a library by identifying its protein product immunologically. Or expressed in another way, if an antibody can be found that is directed against the protein product of the particular gene being sought, the DNA specifying that gene can often be identified.

Antibodies

What are antibodies? And how are they obtained? While it's beyond the scope of this book to go into details on either of these questions, the following is a very brief description. (This section will be referred to during the discussion of recombinant antibodies in Chapter 11.) Antibodies are a class of proteins that are manufactured in vertebrates as a response to invasion by disease organisms. As such, they are able to recognize and bind with great specificity to particular foreign molecules called **antigens**. In order to get antibodies that recognize specific proteins, that purified protein is injected into a rabbit (or some other laboratory animal). If everything goes right, the animal responds to the antigen by synthesizing a burst of antibody molecules which bind specifically and tightly to the protein that had been injected. Because the antibodies have this property, they may be labelled and used to localize a specific protein from a colony that is manufacturing that protein (the principle is the same as that of nucleic acid probes).

Vectors for antibody screening

Obviously, an immunochemical search for an antigen requires that the piece of DNA being sought is capable of coding for a protein. Cloned cDNA's, because they derive from mRNA's, are particularly appropriate for this purpose. In addition, because the protein must be expressed, the vector into which the DNA is cloned must have all the regulatory signals that allow the inserted gene to synthesize a protein. Some of the vectors that are used for this purpose are plasmids, and others are bacteriophage lambda derivatives. A diagram is of one of the plasmids is shown below.

Like most of the other vectors that have been discussed, the DNA to be cloned in these **open reading frame vectors** is inserted at a specific site. (A *Sma*I site is illustrated.) On the left side of this site, the vector bears an initiator region that carries several sequences that encourage efficient transcription

and translation of the passenger from within the initiator region. On the other side of the initiator site, the vector bears most of the bacterial β-galactosidase gene (*lacZ*). The figure at the right illustrates what happens if a passenger is not

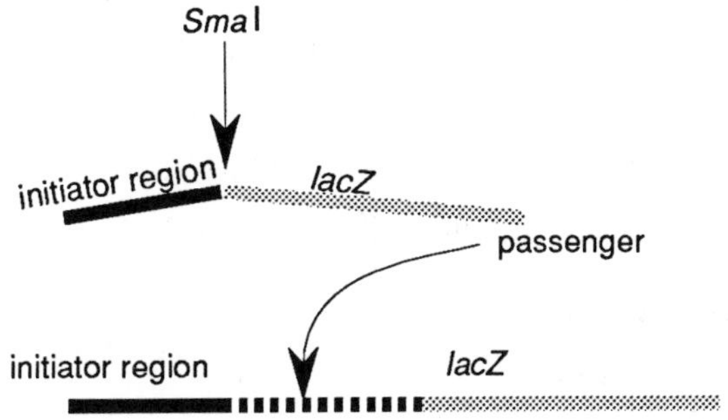

inserted at the *Sma*I site.
Transcription is efficiently
initiated, and a mRNA is
made. Translation of the
mRNA begins in the
initiation region and
proceeds through to sequences within the
β-galactosidase gene. But the translation is
out of phase; the translational machinery is in
the wrong reading frame. That is, after the initial codon is read, translation of
the codons in the β-galactosidase gene don't make sense because the nucleotide
sequence is being read beginning at the wrong base.

Now look what happens if a passenger is inserted. **If** a piece of foreign DNA is
placed at the indicated site, and **if**
this DNA bears a
sequence that can
code for a protein,
and **if** the foreign
DNA just happens
to be in the proper
translational
reading frame, then
proteins started at the ATG will be
synthesized, and synthesis will continue
through the passenger into the *lacZ* gene. What results is a hybrid protein
containing parts of two different proteins – the protein encoded by the passenger
gene and the bacterial β-galactosidase. An essential ingredient of this scheme is
a property of β-galactosidase, the first 25 amino acids of which are not essential
for its activity. Not only that, but these amino acids can be replaced by virtually
any polypeptide, and the resultant hybrid protein will usually have β-
galactosidase activity. The result is that some proportion of the colonies bearing
an insert can be recognized by their ability to stain with a suitable substrate for
the enzyme, β-galactosidase. Moreover, these same colonies can be assayed with
labelled antibody for the presence of the proper antigen.

The lambda GT11 vectors work in a somewhat similar manner, with the exception that the insertion of a passenger inactivates β-galactosidase. In these vectors, advantage is taken of a unique *EcoRI* site near the C-terminal coding region of a *lacZ* gene that is inserted into the phage. If a piece of passenger DNA is inserted here, β-galactosidase is inactivated, and a hybrid protein results.

In either case, the hybrid (or fusion) protein that results is detected with a labelled antibody probe. The details of how this is accomplished are not really important. Suffice it to say that some label can be attached to the antibody, and when it binds to the antigen produced by a colony or plaque, the clone manufacturing that protein can be recognized and eventually isolated.

Summary

Nucleic acid probes can be labelled and used to detect complementary sequences in cDNA or genomic libraries utilizing colony or plaque hybridization.

Probes may be labelled internally by nick translation. Alternatively, DNA molecules can be labelled at their ends by a variety of procedures.

Homologous genes may be used as probes in order to find similar sequences in other organisms. Alternatively, cDNA clones or back translated oligonucleotides can be employed to find a particular sequence within a library.

Specific antibodies can be used to screen expression libraries.

10

Characterization of the passenger

We've covered a lot of ground. A gene from some organism has been introduced into a vector and the specific colony carrying it has been detected with either a homologous probe, with an antibody, by a functional assay, or by some other method. The next task is to begin analysis and characterization of the passenger to make sure that the right gene has been cloned, and to learn something more about its properties.

Restriction mapping

One of the first steps in the characterization of a passenger molecule is to determine the number and position of the restriction sites within it. This procedure is referred to as **restriction mapping**.

The best way of constructing a restriction map is to determine the sequence of the DNA and then to use a computer to search for and identify the position of every restriction site within that sequence. DNA sequencing is covered later in this chapter. But, while DNA sequencing is a widely practiced technique, it remains difficult to do routinely on large pieces of DNA. For DNA's greater than about 5 to 10kb, other methods of restriction mapping must come into play.

In carrying out restriction mapping, the first thing to do is to decide which enzymes to use. As of 1989, more than 1100 different restriction endonucleases had been identified. It would be impractical to position all of their cutting sites within any given piece of DNA. In fact most restriction maps use only about 5 to 15 enzymes – often common ones that have a six-base-pair recognition sequence. For the map that follows,two well-known six-cutters were chosen, but the principles of map construction are the same for any number of enzymes that might be picked.

Two enzymes – sequential and simultaneous digestion

In the following example, the six-base-recognition enzymes *Eco*RI and *Hin*dIII are used to map a 6kb pair linear fragment of DNA. The figure below represents an agarose gel in which the digested DNA has been subjected to electrophoresis.

When the 6 kb pair fragment is cut individually with *Eco*RI and *Hin*dIII, the following results are obtained:

As shown in the left lane of the accompanying figure, *Eco*RI cuts the DNA into four pieces of the indicated size (in kilobase pairs), and *Hin*dIII produces three fragments. The problem in restriction mapping is to determine the location of the sites for each enzyme in the original segment.

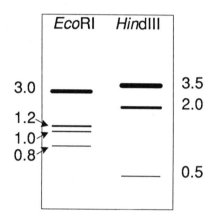

Several pieces of information come from inspection of the data from these digests. First, it should be evident that the number of sites is one less than the number of pieces produced (remember, this was a linear piece of DNA to begin with). So there must have been three *Eco*RI sites and two *Hin*dIII sites in the original 6 kb piece of DNA.

Second, it also should be clear that there are many possible positions that the sites could occupy and that there is insufficient information from the simple single digests to decide among them. In order to position the sites within the original fragment, one strategy is to digest with the two enzymes sequentially. In the adjoining diagram, each of the four *Eco*RI pieces has been isolated and

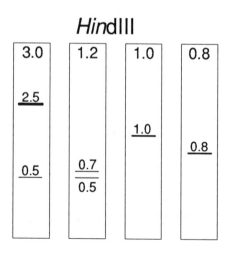

subjected to digestion with *Hin*dIII. A similar analysis can often be done by simply carrying out the two digestions simultaneously, although the results may be complicated by the presence of many bands, some of which may not be resolvable from one another.

Inspection of the digest shows that two of the original *Eco*RI fragments were further digestible with *Hin*dIII. It's easy to tell which *Eco*RI pieces contained *Hin*dIII sites because pieces of the original size are no longer found, and new pieces of smaller size replace them.

Now there is enough information to make a rudimentary map. The idea is to use logic and trial and error in solving what amounts to a minor topological puzzle. For example, there are 24 ways that the four different *Eco*RI fragments could be arranged. So, one strategy is to draw all 24 arrangements and then see what would result if each were subject to *Hin*dIII digestion.

As an example, suppose the four sites were arranged as in the figure at the right. It is known from the double digests that the 3kb piece and the 1.2kb piece have *Hin*dIII sites within them. No matter how the location of the *Hin*dIII sites are arranged (again assuming that this is the arrangement of the *Eco*RI sites), it is impossible to get the pieces that were obtained in the *Hin*dIII digestion alone. This must mean that the initial provisional arrangement of the four *Eco*RI fragments must have been wrong. By choosing other possible orderings of the *Eco*RI fragments, it should be possible to arrive at an arrangement of *Eco*RI pieces that yields a map consistent with the *Hin*dIII results.

0.8

3.0   1.2   1.0

Possible arrangement of the *Eco*RI sites

End labelling

Because deriving map positions from double digests is like solving a puzzle, it can be fun (but time consuming) to do restriction mapping this way. For those less inclined to games, other simpler techniques can be used. These may be less enjoyable intellectually, but they yield faster results. However, they do require some additional laboratory manipulations. Perhaps the most straightforward of these methods is to digest DNA that is end labelled. Here, one end of the fragment is labelled with a radioactive phosphate and incompletely digested with the appropriate restriction enzymes. This incomplete or partial digestion results in fragments that are not cut in all their sites. After the digestion, the resulting fragments are electrophoresed.

Partial digestion might be expected to yield a very complicated mix of pieces. And it does. But the end labelling saves the day. The gel is autoradiographed, and because only those pieces of DNA with a labelled end show up (they are the solid lines in the figure at the right), the pattern of banding is relatively simple. The number and arrangement of the sites is simply read off the autoradiograph. Another view of the fragments is shown in the figure below. The lines with circles at the end represent pieces of DNA whose ends are labelled.

———	6.0
------	5.0 4.8
———	4.0 3.8

------	3.0 **kb**
------	2.0
------	1.2 1.0
------	0.8

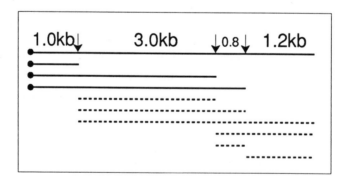

Chromosome walks

As we've seen, genomic libraries consist of collections of more or less random DNA fragments cloned in a particular vector. Some of these fragments will overlap. In fact, if the library is extensive enough, almost every clone will have an overlap with some other. Advantage can be taken of these overlaps to "walk" along a chromosome from some cloned site to another located nearby.

To begin a chromosome walk, a cloned piece of passenger DNA must first be restriction mapped. There will be one restriction fragment near the very end of the cloned piece. (Of course, there will be one at each end. Either end can be selected, depending on the direction of the walk.) An end fragment (marked "11" in the figure) may be isolated, labelled, and used as a hybridization probe. The probe, in turn, is used to identify a set of clones from a genomic library. All members of the set of clones that hybridize with the probe will contain a passenger that shares some sequences with the probe.

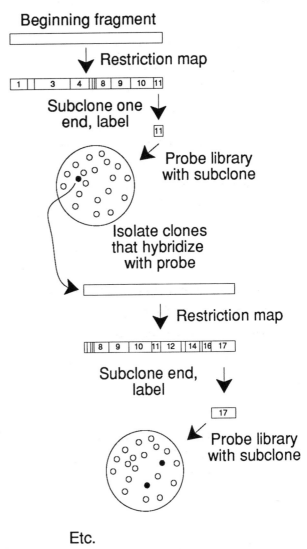

Beginning fragment

↓ Restriction map

| 1 | | 3 | 4 | || | 8 | 9 | 10 | 11 |

Subclone one end, label ↓

11

↖ Probe library with subclone

Isolate clones that hybridize with probe

↓ Restriction map

| || | 8 | 9 | 10 | 11 | 12 | | 14 | 16 | 17 |

Subclone end, label ↓

17

↖ Probe library with subclone

Etc.

Some members may include new sequences that extend out in one or both directions. Restriction mapping of these new clones will reveal the extent of overlap and in which direction unmapped sequences extend. Restriction mapping may also reveal a new end fragment (marked "17") that can serve to probe additional clones. By repeating this process, a series of overlapping clones can be collected that cover large regions of a chromosome. In practice, areas encompassing hundreds of thousands of base pairs of contiguous DNA have been cloned from a variety of organisms.

Chromosome walking has proved to be a powerful technique for characterizing the DNA that borders a piece of cloned DNA. It also can be used to clone a gene that is known from genetic data to lie close to a DNA species that has already been cloned.

Southern blotting

As shown above, recombinant DNA techniques are useful for purifying and characterizing particular pieces of DNA. The process can be often turned around, and a cloned DNA can be used to learn more about the genome from which it came without having to resort to further cloning. For example, a cDNA clone can be used to find out more about its matching genomic DNA. Or a genomic clone can be used to investigate the area in and around similar genes in genomes from closely related organisms (See "RFLP's" in the next chapter.)

Southern blotting allows investigators to do restriction maps on uncloned genomic DNA using a cloned probe. The name of the technique derives from Edward Southern, who originated it in 1975. In a typical Southern blot, DNA is extracted from an organism, treated with one or more restriction endonucleases, and then subjected to gel electrophoresis. After electrophoresis, the DNA is transferred to a nitrocellulose filter – this is the blotting step – where it can

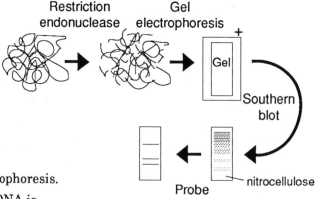

be visualized by virtue of its hybridization to a labelled probe. In principle, the technique is similar to colony hybridization in that the idea is to transfer the DNA from the gel to the filter, retaining the position of the individual fragments. Some specific uses of Southern blotting are illustrated in the next chapter.

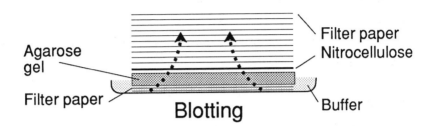

Blotting

Some years after Southern's technique came into general use, a variant of the procedure involving the transfer of RNA to an insoluble support was given the name "Northern blotting" even though the name of the author of the new technique had nothing to do with direction. Scientists, apparently without embarrassment, then took this game even further and gave the name "Western blotting" to the transfer of electrophoresed proteins to filters.

Sequencing

With the advent of two procedures for nucleic acid sequencing (**Maxam/Gilbert** and **Sanger** procedures), it is now possible to sequence thousands of nucleotides daily. New machines are being developed that may increase this number by an order of magnitude or more. Once the sequence of a gene is known, a good deal of useful information becomes available. For example, a genetic engineer can easily determine the number and location of all the restriction enzyme sites without recourse to experimentation. In addition, if there is doubt that the right gene has been cloned, translation of the complete sequence (a job best left up to the computer) and comparison with the amino acid sequence of the protein can ensure that a mistake hasn't been made.

The Sanger procedure

The two sequencing techniques differ considerably, but both make use of high-resolution polyacrylamide gel electrophoresis to separate a series of DNA molecules differing by as little as a single base pair in size. The Sanger (or dideoxy) procedure is easier to understand and is more widely used. It utilizes DNA synthesis with a single-stranded DNA (the DNA to be sequenced) as a template. The basic idea is to start synthesis at a specific place on the template and terminate the growing chains at defined nucleotides – either A, C, G, or T.

The first goal is reached by beginning synthesis from a short oligonucleotide – a **primer** – that is complementary to a specific sequence

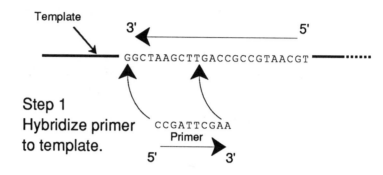

on the template. During synthesis, radioactive nucleotides are incorporated into the growing chain. Thus all chains begin at a specific place and are radioactively labelled.

Termination is achieved by the incorporation of a chain-terminating dideoxynucleotide. Dideoxynucleotides lack a 3' OH group and therefore cannot couple with the next, incoming deoxyribonucleotide. Thus wherever a dideoxynucleotide is incorporated, the growing chain terminates. If one adds some dideoxy A, for instance, into a reaction mix, each time that it is incorporated, the growth of the chain will stop. This results in a molecule with a dideoxy A at the terminus of a piece of DNA where ordinarily an A would go. But since the synthesis mixture contains deoxy A's as well as dideoxy A's, synthesis will

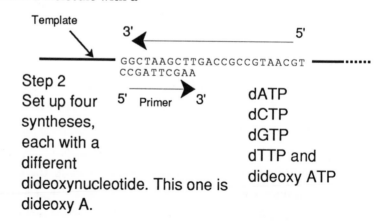

terminate across from different T's, depending on where in the chain a dideoxy A happens to be incorporated. Thus a series of molecules, each terminating with a dideoxy A, will be generated. If this mixture is electro-phoresed, a "ladder" of

Step 3
Termination occurs across from this T
or this one,
or even further downstream,
depending on whether a dideoxy A or deoxy A
is incorporated.

```
———————— GGCTAAGCTTGACCGCCGTAACGT ———……
                CCGATTCGAACTGGCGGCA
Product 1 ——→ CCGATTCGAACTGGCGGCATTGCA
Product 2
```

molecules will be displayed. The position of each rung corresponds to where an A is located. By doing separate reactions with dideoxy T's, G's, and C's, one can tell where in the chain each of these nucleotides belongs.

The Maxam/Gilbert procedure

In the Maxam/Gilbert method, DNA is end labelled and partially cleaved at each of the four bases in four separate reactions. The resultant pieces are separated by acrylamide electrophoresis, and the sequence is read off an autoradiograph by determining which base-specific agent cleaved at each nucleotide.

Other techniques

In situ hybridization

Under some circumstances, *in situ* hybridization can be a powerful technique for characterizing some genes. In it, a labelled probe is hybridized to either cell squashes or histological sections. If the chromosomal location of a particular gene is known (by ordinary genetic analysis, for

Hybridized probe (enlarged 1000 times for clarity)

←——— 100,000,000 bp ———→

example), this kind of experiment can confirm that the gene in hand is the desired one. Alternatively, *in situ* hybridization can be used to determine the location of a cloned gene on a specific chromosome.

Because of limitations in sensitivity of the technique, *in situ* hybridization was first used for the characterization of genes that occur in clusters of more than one copy. For example, in most organisms there are often hundreds of genes coding for ribosomal RNA. They are often found together in the chromosome and a single probe will pick up the entire cluster. More recently, sensitive fluorescent probes have made it possible to localize genes on mammalian chromosomes that occur as a single copy, as indicated in the figure on the previous page.

Hybrid selection and translation

Another generally applicable technique for characterizing a passenger DNA is called **hybrid selection and translation** (sometimes the term **hybrid released translation** is used). In this procedure, a cloned DNA is bound to some insoluble support and used to "fish" out a specific mRNA molecule from the total RNA of a cell or tissue by nucleic acid hybridization. That is, only those RNA's that are complementary to one strand of the bound DNA will hybridize with and bind to it. The unhybridized RNA is then washed away, leaving only those species that have been captured. These RNA's are then released (by denaturation) and placed in an extract that is capable of translating mRNA into a protein. After translation, a protein is produced. Its sequence, of course, is determined by the sequence of the mRNA's that were caught during the fishing. And the captured RNA's must have been complementary to the cloned passenger DNA.

The final result of all of this is that the protein that is produced reflects the gene that is cloned. The protein can be identified by its reaction with specific antibodies. Or its molecular weight can be established by electrophoresis. It can even be sequenced. In the end, the characteristics of the protein can be used to confirm the identity of a cloned passenger.

Hybrid-selected translation can also be used to screen tens or perhaps hundreds of cloned DNA's for the appearance of a specific protein. The technique is not particularly suited for screening a library, but it is very useful for testing a small group of colonies in which a suspected cloned gene of interest lies.

Summary

Cloned DNA may be subjected to restriction enzyme mapping, sequencing, *in situ* hybridization, or hybrid selection and translation. These techniques can confirm the identity of a particular clone and extend our understanding of its properties.

11

Applications of gene cloning

Genetic engineering promises to revolutionize medicine and agriculture and to have a major impact on the chemical and pharmacological industries. The new procedures and products that it will engender seem certain to have a growing influence on our lives in the coming decades.

Diagnosis

Several new diagnostic methods have originated from recombinant DNA technology. The diagnosis of sickle cell anemia serves as a convenient illustration of some of the tools that recombinant DNA has made available and of the progress that has been made over the years. The ethical dilemmas that come with this new technology are covered in the last chapter.

Sickle cell anemia

Sickle cell anemia is a genetic disease with the distinction of being the first such condition whose basis was worked out at the molecular level. Sufferers from sickle cell anemia differ from normal individuals in that instead of carrying the normal β^A gene (which codes for the β-chain of hemoglobin), they carry the β^S variant. This mutation results in the synthesis of a β-globin that bears the amino acid valine instead of glutamic acid in position #6. People who are homozygous for this mutation (that is, both homologous chromosomes carry the mutation), have red blood cells that assume an

Hemoglobin A: Val-His-Leu-Thr-Pro-Glu-Glu-...

Hemoglobin S: Val-His-Leu-Thr-Pro-Val-Glu-...
                  1   2   3   4   5   6   7

abnormal shape (sickling) under low oxygen pressure. Such sickled cells drastically impede circulation, and most homozygotes do not survive to adults.

The β-globin gene was the first human protein-coding gene isolated. Once the gene was cloned, portions could be labelled. When these were used to probe Southern blots, it was discovered that individuals with sickle cell anemia often differ from people without the disorder in a *Hpa*I restriction fragment. Normal individuals usually have a 7kb *Hpa*I restriction fragment that hybridizes with a hemoglobin probe; sickle cell anemics often lack a *Hpa*I site and consequently have a 13kb fragment instead (lane marked "fetus" in figure).

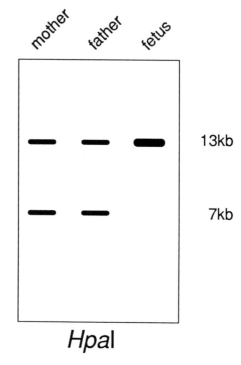

Notice that the mother and father of the affected fetus have both 7kb and 13kb fragments, indicating that they are heterozygous for the sickle cell gene. A homozygous normal individual would only show a 7kb fragment.

RFLP's

Two important points merit emphasis. First, not everyone with the 13kb fragment carries the gene for sickle cell anemia, and not everyone without it bears only normal genes. The fact that only a good correlation is observed between the presence of the abnormal gene and the 13kb fragment is a strong indication that the change in DNA that caused the restriction length difference wasn't responsible for the sickle cell mutation. In fact, it was discovered later that the *Hpa*I site found in most normal individuals is located more than 5kb away from the β-globin structural gene, and its loss has nothing to do with the disease. By chance, the disease must have originated in someone who was missing this *Hpa*I site.

Second, there are many such restriction site differences all over the genome in the human population. In fact, it has been estimated that unrelated humans differ in as many as 0.3% of their DNA positions. That is, if someone were to sequence a particular 1000 base pair segment of DNA from two unrelated individuals, they would find, on the average, three differences in DNA sequence between them. As you might expect, these differences sometimes cause a loss or gain of a restriction site and consequently can result in diverse sizes of restriction fragments in an assortment of individuals. Differences between the size of restriction fragments in a population are called **restriction fragment length polymorphisms** or **RFLP's** (pronounced RIFF-LIPS). As demonstrated by the example above, the presence of RFLP's can serve as a powerful diagnostic tool in the analysis of genetic diseases.

It was soon recognized that the mutation that causes the amino acid difference that distinguishes most sickle cell anemics from normals is a single base change

Hemoglobin A: Val-His-Leu-Thr-Pro-Glu-Glu-...
                              CCT GAG GAG

Hemoglobin S: Val-His-Leu-Thr-Pro-Val-Glu-...
                              CCT GTG GAG
                                       A to T

in the DNA sequence, CCTGAGG. This sequence codes for proline (CCT), glutamic acid (GAG), and the first G in the code word for another glutamic acid (see the figure above). The disease is caused by an A to T change that results in the new sequence CCTGTGG, thereby transforming a glutamic acid (GAG) codon into a valine (GTG). It happens that the restriction enzyme *Mst*II cuts at the sequence CCTNAGG and therefore cleaves normal DNA in the β-globin gene but fails to cut the gene of a sickle cell mutant. As a result, if human DNA is cut with *Mst*II, a difference will be seen on Southern blots between sickle cell individuals and normals. Notice that the diagnosis with *Mst*II is fundamentally different from the very similar one done with *Hpa*I. In the *Mst*II example, the change in sequence that is being probed is the same one that caused the disease. It is not simply a correlation. The molecular basis of the disease is actually being used in its diagnosis.

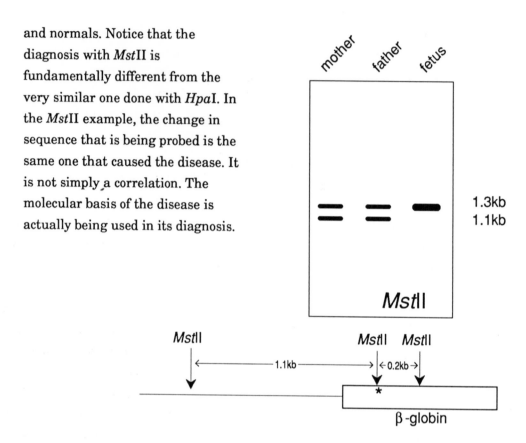

Allele-specific probes

A further refinement was necessitated when it was found that certain populations carried another mutation, a change from CCTGAGG to CCTAAGG, that resulted in another anemia (this variant is called hemoglobin C and occurs in high frequency in West Africa). Hemoglobin C substitutes lysine (AAG) for glutamic acid (GAG). The sequence in this mutant can be cut by *Mst*II, and a Southern blot done after the use of this enzyme is not diagnostic.

Hemoglobin A: Val-His-Leu-Thr-Pro-Glu-Glu-...
                              CCT GAG GAG

Hemoglobin S: Val-His-Leu-Thr-Pro-Val-Glu-...
                              CCT GTG GAG
                                      A to T

Hemoglobin C: Val-His-Leu-Thr-Pro-Lys-Glu-...
                              CCT AAG GAG
                                      G to A

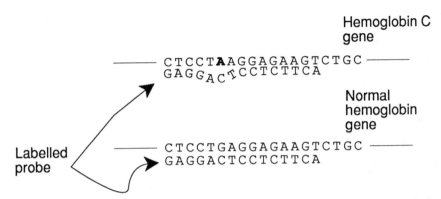

In order to be able to distinguish between the two mutant forms and the normal one, molecular geneticists resorted to the use of **allele-specific** (sometimes called sequence-specific) **oligonucleotide probes**. For the diagnosis of anemia due to hemoglobin C, an oligonucleotide probe is synthesized that is complementary to the hemoglobin C gene at the site at which the mutation has occurred. The probe is labelled using one of the techniques previously described. DNA is then extracted from the individual to be tested and transferred to nitrocellulose filters. The filters are exposed to the probe under conditions where only perfectly complementary hybrids will form. If there is a difference of even a single base between the probe and the DNA on the filter, the probe will not be retained and there will be no signal detected. If DNA from a hemoglobin C homozygote is placed on the filter, it will bind and retain the probe. If DNA from a normal individual is put on the filter, hybridization will be imperfect, and a signal will not be detected. If DNA from a hemoglobin C heterozygote (with one normal β-globin gene and another bearing the β^C mutation) is placed on the filter, the probe will only show about half the normal signal. The hybridization is shown diagrammatically in the figure above.

Sequence-specific oligonucleotide probes work well for many purposes, but they have their limitations. For example, it is known that mutations in other regions of the β-globin gene can cause blood disease. To diagnose these disorders, other probes would have to be fashioned. In fact, a different probe would have to be synthesized for each region of the gene that was to be examined. DNA sequencing of the entire gene – the ultimate diagnostic tool – would avoid this problem because it would reveal all possible changes. But in order to sequence a gene, it is usually necessary that it be cloned. That means that the hemoglobin gene from each individual to be tested would have to be cloned; a relatively laborious procedure requiring a fairly large amount of tissue as starting material. Diagnosticians began to look to an alternative to cloning and sequencing. And they soon found one.

PCR

Recently a new technique has been developed that allows scientists to analyze DNA from an extremely small amount of starting material without having to resort to repeated cloning. It has even allowed sequence analysis of DNA from a single human hair root! The technique is called **polymerase chain reaction** (**PCR**). The principle behind PCR is simple. A specific fragment of DNA is repeatedly synthesized using the enzyme DNA polymerase. The result is the amplification of a particular sequence – by a millionfold or more.

The method works as follows: Two primers are synthesized that flank the region that is to be amplified. (They are shown as short, thick lines with arrowheads in the figure.) Total DNA is extracted and purified from the sample under study. The DNA is then denatured and the primers added in excess and hybridized to the DNA. DNA polymerase is then added, and DNA synthesis begins. Two newly synthesized molecules (indicated by dotted lines in the figure) result. They begin at the primers and extend indefinitely in one direction.

The DNA is then heat denatured again, more enzyme is added (additional primer is not needed because it was added in excess in the first step), and the process is repeated. At the second round of synthesis, some primers hybridize to the molecules that

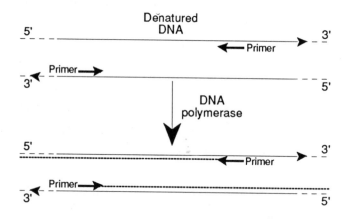

had been synthesized in the previous round and produce a new class of DNA that extends from one primer to the other. After each round of synthesis, the amount of this class of DNA increases geometric-ally so that after 25 rounds the region between the two primers can increase some million times or more.

One drawback of the original PCR method was that additional DNA polymerase had to be added after each heating step because the high temperatures needed to separate the DNA strands denatured the enzyme. In 1988, a thermostable DNA polymerase obtained from an organism that lives in hot springs (*Thermus aquaticus*) was substituted for the corresponding *E. coli* enzyme. It is capable of enduring exposure to 95° C and didn't need to be replenished after each cycle. With the advent of this new enzyme, a sample containing primers and template is simply repeatedly heated and cooled. During the heating cycle, the DNA is denatured. When the reaction is cooled, the primers hybridize, and new strands are synthesized.

PCR has several limitations. First, sequence-specific primers are required, and that means that the DNA sequence of the region flanking the amplified DNA must be known. In turn, this means that somewhere along the line, the gene being amplified must have been cloned and at least partially sequenced. But it is important to realize that once a gene has been sequenced, PCR can be used to amplify the DNA from any number of individuals, including potential mutants. In other words, for diagnostic purposes the cloning only has to be done once. A second drawback of PCR is that only a limited stretch of DNA can be amplified. The distance between the two primers is restricted to something less than 5kb. A third problem with the method is that the polymerases used for DNA replication make occasional mistakes. If such errors occur early in the reaction, they may be amplified along with the correct sequences, and incorrect diagnoses may result.

DNA amplified by PCR can be immobilized on filters and probed with allele-specific oligonucleotides. But many human mutations, especially dominant and X- linked ones, are eliminated rapidly from the population because of their severe detrimental effects. That means that for many diseases, each affected family probably bears an independent mutation. PCR has the ability to solve this problem because PCR-amplified DNA can be sequenced, and any kind of mutations that occur between the flanking primers can be revealed.

Products of gene cloning

One of the initial attractions of recombinant DNA technology was that it appeared that nearly any foreign protein could be synthesized in bacteria or yeast in virtually unlimited quantity. The pharmaceutical industry jumped at this opportunity. They quickly realized that there were many enzymes, non-enzymatic proteins, and peptides that might be commercially viable products if they could be had in abundance. During the past 15 years, these initial expectations have been met, and a new industry based on genetic engineering has arisen.

Peptides

There are a myriad of peptides found in humans and other animals that serve important biological roles. Generally, they fall into three categories: **growth factors** (like insulin, growth hormones, and erythropoietin), **neuroendocrine peptides** (enkephalins and endorphins are two examples), and **glycoprotein hormones** (such as follicle-stimulating hormone and luteinizing hormone). Genetic engineering has made a contribution to the production of products in each of these categories. Insulin is a good example.

Insulin

Human insulin was one of the first large-scale products of genetic engineering. Its initial route of synthesis (many variations now exist) is instructive because it illustrates many of the problems often encountered in trying to produce a mammalian protein in bacteria. In order to understand the rationale for the steps taken, a bit of insulin biochemistry must be understood. To begin with, the insulin structural gene is 1430 nucleotides long. It is interrupted by two intervening sequences of 179 and 786 nucleotides. The gene encodes a protein that is 110 amino acids in length. But the mature insulin peptide looks quite unlike the protein that would be expected from this gene arrangement. It consists of two chains, called A and B, held together by bonds formed between the amino acid cysteine on adjacent chains (marked S in the figure). The A chain is only 21 amino acids long, and the B chain is only 30. Something must have occurred between the synthesis of the 110-amino-acid protein and the formation of the much smaller active molecule composed of two chains. That something was protein processing.

Insulin is processed in two steps. The primary product is called **preproinsulin** and, as predicted from the DNA sequence, is indeed 110 amino acids in length. In the first step of processing, the "pre" part of the insulin is lost during its secretion through the cell membrane. The part that is lost, the first

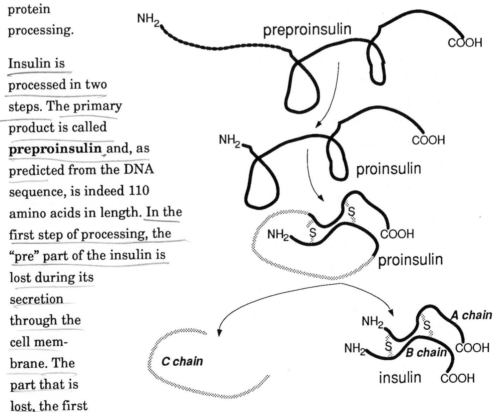

24 amino acids, serves as a **signal** peptide and helps to guide the protein through the membrane. What remains is an 86-amino-acid protein called **proinsulin.** It is converted to insulin inside vesicles within the cells of the pancreas. An enzymatic cleavage takes place that removes an internal fragment of 33 amino acids (called the C or connecting chain) and a few assorted residues at its ends, leaving the A and B chains (with 21 and 30 amino acid residues, respectively). These associate with each other to form mature, biologically active insulin.

It wouldn't have made much sense to clone the insulin gene and use it to produce the final product. After all, the presence of the intervening sequence and the "pre" and "pro" parts would complicate matters considerably. In fact, as discussed in Chapter 8, bacteria will not remove intervening sequences: not from insulin, nor from any other eukaryotic RNA. Furthermore, the specific set of peptide-processing steps that insulin undergoes won't occur in bacteria either. So the decision was made to chemically synthesize the genes for each of the two chains, rather than try to obtain them some other way.

The overall strategy was to produce two bacterial strains, each carrying a different insulin chain. The decision was also made to append the insulin chains to a preexisting bacterial gene: β-galactosidase. In that way, the amount of the hybrid protein could be conveniently regulated (because the amount of β-galactosidase can be manipulated by adding substances that induce its synthesis). Finally, the linkage between the insulin chains and β-galactosidase was made via a

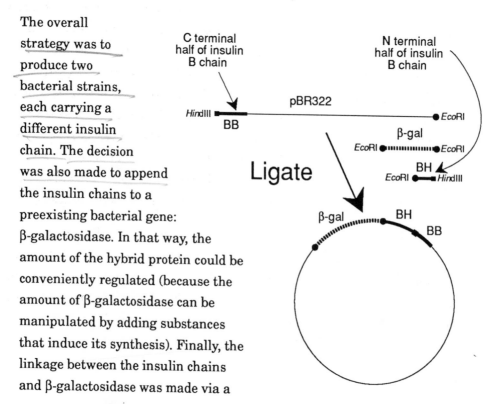

methionine residue. It turns out that there is a classic chemical treatment that cleaves proteins at methionine residues. If the peptide is linked to β-galactosidase via methionine, it can be conveniently freed (there are no methionine residues in either of the insulin chains) and ultimately purified from the β-galactosidase. In the final step, each of the insulin peptides would be purified and then mixed together under conditions where the dimer would form.

The details of the construction are shown in the accompanying diagram. The B chain was constructed in two parts. First, eight overlapping deoxyoligo-nucleotides were ligated together to form the amino terminal part of the coding sequence for the B peptide (called BH). The first and eighth oligonucleotides were constructed so that they bore EcoRI and HindIII cohesive ends, thus allowing ready insertion into the cloning vehicle, pBR322, that had been cut with the corresponding enzymes. The C terminus, the BB fragment, was synthesized in a similar manner from ten deoxyoligonucleotides. The first and tenth oligonucleotides bore HindIII and BamHI ends. After ligation, the resulting BB fragment was inserted into pBR322 that had been cut with HindIII and BamHI. The two plasmids, each bearing one half of the B chain, were cut with EcoRI and HindIII, thereby liberating the two insulin gene-containing passenger fragments shown in the figure. These were purified and mixed with a third piece of DNA (a fragment containing the regulatory region and most of the coding region of the β-galactosidase gene) that had been cut with EcoRI, and the whole jumble ligated together (see diagram). When the N terminal coding side of the insulin B gene is ligated to the C terminal coding end of the β-galactosidase structural gene, a hybrid gene is formed that codes for a protein that is mostly β-galacto-sidase with the insulin B chain appended to its C terminus. A similar, but simpler, procedure was used to produce the insulin A/β-galactosidase fusion.

Bacterial strains containing the gene for either the A or B chain of insulin attached to β-galactosidase produced large amounts (about 20%) of their total protein as hybrid protein. After chemical treatment to release individual insulin chains away from the β-galactosidase, it was found that insulin activity could be reconstituted by simply mixing together the purified individual chains.

Proteins

Some proteins (as well as many of the peptides mentioned above) must be modified before they become biologically active. For instance, the example that we've just encountered, preproinsulin, must be cut in several places before it can become an active insulin molecule. With insulin, it was possible to mimic these modifications by shortening the insulin gene and breaking it into two parts before inserting it into a bacterial vector. But proteins undergo subtler changes that are impossible to simulate by changing the sequence of the corresponding gene. For example, many proteins have carbohydrate groups bound to certain amino acids. Others may carry phosphate groups. These and other modifications are imposed on the proteins after they are synthesized. Specific enzymes add chemical groups onto specific residues of the completed protein. Often these modifications are required for the activity of the targeted proteins.

The point of this discussion is that some of the enzymes that modify proteins may be present only in the organism in which that protein is normally made. If the synthesis of a protein is forced to occur in bacteria or yeast, some modifications may not occur at all. Alternatively, the protein may be modified incorrectly. This situation poses a dilemma for the genetic engineer who wants to produce a particular protein in high yield in a convenient host. However, new hosts and vectors have allowed imaginative answers to this problem. One such solution is illustrated: the use of insect cells in culture to make the lymphokine, **interleukin-2**.

Lymphokines

Lymphocytes are a kind of leukocyte or white blood cell. Some classes of lymphocytes secrete antibodies. Others respond to antigen by secreting substances that interact with other white blood cells. These substances are called **lymphokines** or **interleukins**. They include such proteins as the **interferons** (α, β, and γ), **interleukin -2, -3,** and **-4, macrophage migration inhibition factor (MIF)**, and **tumor necrosis factor** (TNF). All of these proteins play a vital role in the body's cellular defense against disease. For example, MIF activates macrophages (another kind of white blood cell) so that they become more efficient in attacking and engulfing invading microorganisms. Interleukin- 2, a lymphokine that is sometimes called **T-cell growth factor,** stimulates the propagation of activated T cells, cells that are important in the antigenic response to infection that is elicited by invading viruses and fungi.

Prior to the recombinant DNA revolution, most lymphokines were only obtainable in minute quantities, but it did not take much imagination to see that they might be valuable therapeutic aids if their genes were cloned and they were made more readily available. The interferons were particularly attractive in this regard, and their genes were indeed cloned, and the manufacture of recombinant interferon is currently a big business. Interleukin-2 is another lymphokine whose gene has been cloned. Its production in insect cells offers some insights into the considerations involved in choosing a suitable host.

While interleukin-2 has been produced in *E. coli* and even in mammalian cells, the synthesis of the protein in insect cells offers several advantages.

First, insect cells are easy to cultivate. Moth and butterfly cells can be grown in culture at lower temperatures than mammalian cells, and they do not require high concentrations of CO_2 (as do mammalian cells).

Second, insect cells can be infected with a virus (a **baculovirus**) that can serve as a convenient vector for foreign genes. The result can be the production of prodigious amounts of recombinant DNA products.

Third, unlike bacterial systems, but like mammalian and yeast cells, insect cells will carry out most posttranslational modifications that certain eukaryotic proteins require. Insect cells, for example, will correctly remove the signal peptides that tell a protein that it is to be secreted. They will not, however, properly transfer most of the carbohydrate residues that are found in mammalian proteins because they lack some of the enzymes found in vertebrates.

Last, baculoviruses are not pathogenic to vertebrates or plants, an important consideration for companies contemplating large-scale commercial distribution.

To get synthesis of interleukin-2 in insect cells, it was first necessary to introduce the gene into a baculovirus vector. The most widely used viral vector was first isolated from a moth called the California looper, *Autographa californica*. The virus, commonly called AcNPV (for *Autographa californica* nuclear polyhedrosis virus), has a large (128kb), double-stranded, circular genome. One of the particular characteristics of the baculoviruses is that they form prominent polyhedral crystals within the nuclei of the cells that they infect. The crystals consist of numerous virus particles enveloped in a gluey matrix of a protein called **polyhedrin**. It is the polyhedrin gene that is the key to success in cloning foreign genes into the polyhedrosis virus.

While the virus ordinarily forms polyhedra in the nuclei of infected cells, the polyhedra are **not** required for the spread of the virus in culture. (The polyhedra are, however, required for the virus to infect whole insects. The crystals are ingested by the insects, and during passage through their digestive system, the polyhedrin glue is broken down thereby releasing virus particles.) Instead, in cell culture, the virus buds off through the plasma membrane to infect other cells. Thus polyhedrin protein is dispensable in cell culture even through it is abundantly synthesized late in infection, becoming the most plentiful protein found in infected cells.

Insect virologists realized early on that if a foreign structural gene could be substituted for polyhedrin, the virus could be fooled into making large amounts of foreign proteins without untoward effects on viral multiplication in culture.

Given these properties, vector construction was begun. However, a complication soon arose. The viral DNA is very large with many restriction sites all over its genome. Moreover the virus was (and remains) mostly unmapped genetically. It was soon realized that it would take many years and much effort to place convenient restriction sites at

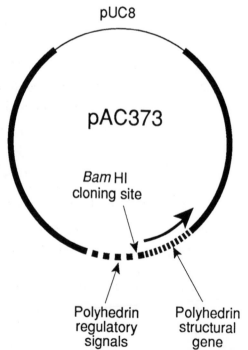

pUC8

pAC373

Bam HI
cloning site

Polyhedrin
regulatory
signals

Polyhedrin
structural
gene

appropriate places within the virus so that foreign genes could be introduced. Instead, another strategy was used. Using the small bacterial plasmid pUC8 as a framework, a series of vectors was constructed. The first step was to introduce a 7kb piece of DNA containing the viral polyhedrin gene and surrounding sequences into the plasmid. Then convenient cloning sites were placed downstream of the regulatory signals of the polyhedrin gene. In one plasmid, pAc373, a *Bam*HI site was positioned appropriately, and into this site a cDNA that coded for the entire interleukin-2 gene was inserted.

The next step was to get the gene from the plasmid into the virus. This was done by exposing insect cells to a calcium phosphate precipitate of a mixture of DNA from the plasmid pAc373 and normal polyhedrosis virus. In about 1% of the cells, the gene on the plasmid undergoes homologous recombination with the corresponding polyhedrin sequence on the virus, thereby inserting foreign DNA into the virus at the polyhedrin gene. (Genetic recombination – an exchange of genetic information – of this sort commonly occurs between homologous DNA sequences in a great variety of organisms. As discussed below, genetic engineers regularly take advantage of the phenomenon to move genes from small, easily manipulable plasmids into larger plasmids and viruses.) This recombinant virus was no longer capable of making polyhedrin and therefore synthesized virus without polyhedra. In fact, the absence of polyhedra served as a convenient screen to distinguish cells harboring the recombinant virus from cells containing viruses that hadn't undergone recombination. The cells could be isolated and grown as pure cultures. They produced large quantities of the recombinant DNA product.

In the specific case of interleukin-2, it was found that the protein was synthesized in considerable quantity. Moreover, the insect cells were found to have secreted interleukin-2 into the medium, indicating that the secretory signal peptide found on the protein was being recognized and properly processed. However, there was no evidence that the correct set of carbohydrate groups had been added.

In summary:

   • Interleukin-2 and other lymphokines are important biomolecules that will undoubtedly become of increasing importance for the treatment of disease in the future.

   • Because of their limited availability from natural sources, their production by genetic engineering is vital if they are to become widely used.

   • Insect cells combined with the baculoviruses have served as convenient hosts for the production of lymphokines.

   • The main advantages of the insect cell system are its high yield of recombinant product and the ability of the cells to carry out many (although not all) of the post-translational processing required to bring the protein to its final form.

In vitro antibody production

There has been a recent advance in the use of genetic engineering to produce antibody molecules, one that promises to revolutionize the ease with which antibodies, particularly human antibodies, are obtained. But in order to logically develop this topic, it will be necessary to recall what an antibody is and how antibodies are produced. This discussion will also serve to introduce the subject of vaccines, a topic that is covered after this one.

As already explained in Chapter 9, antibodies are proteins produced by the immune system in response to exposure of the body to a foreign macromolecule. The foreign substance is called an **antigen.** Carbohydrates and proteins are good antigens, meaning that if they find their way into the body, animals readily produce antibodies against them. Almost any protein or carbohydrate from any source, no matter how obscure, will elicit antibodies. (However, proteins and carbohydrates from the animal itself are ordinarily not immunogenic. Organisms have a mechanism that recognizes self, thereby preventing autoimmune disease.) Even newly invented chemical groups, when chemically attached to a macromolecule, will cause the production of antibodies. What's more, the antibodies that are produced will specifically and very tightly bind to that group.

What is the mechanism that allows an organism to make an antibody against virtually any foreign antigen that it encounters? Recombinant DNA methods have played a major role in unraveling this important question. The answer, briefly, is that organisms don't design antibodies to order. Instead they synthesize millions and hundreds of millions of different antibodies, each encoded by a different (but related) gene, each one with a different specificity, and each housed in a different cell. Every one of these cells then stands ready, with its particular antibody poised on its surface, waiting for an antigen to arrive to which it can bind. If an appropriate antigen is encountered, that specific cell begins to divide, and its daughters eventually secrete the antibody.

The design of this system seems extraordinarily wasteful. The body makes millions of different antibodies, each with its own specificity, the vast majority of which are kept in reserve and never used because they never encounter an appropriate antigen. However, several resourceful scientists realized that this apparent inefficiency could be put to good use. They reasoned that if the entire collection of antibody-producing genes could be moved into *E. coli*, with a single different antibody gene in each of millions of bacterial cells, then the entire repertoire of potential antibodies could be at their disposal.

To accomplish this feat, a group at the Scripps Institute in California used a variety of genetic engineering methods that have already been discussed. First, they used reverse transcriptase to make a cDNA copy of total antibody mRNA and then employed PCR to specifically amplify the cDNA's. (A critical point to keep in mind is that the PCR used here amplifies many sequences. While each of the antibody genes is different, they do share certain sequences. The primers used for PCR were designed to hybridize to the regions shared by the genes and therefore to amplify each of the cDNA's.) Second, they placed all these antibody gene sequences into a lambda bacteriophage library by inserting the genes into an appropriate site in the vector. Third, they arranged the lambda vector so that an inserted gene would synthesize a protein in bacteria and that this protein would be secreted by the bacterial cells.

The result of this process was a library consisting of about one million clones. Each clone (or at least the great majority of them) expressed an antibody with a unique specificity. To assay whether a specific clone carries an antibody with a specific affinity to a certain antigen, the protein from each plaque was first fixed to a nitrocellulose filter. A solution containing a labelled antigen was then placed on the filter. If it was bound by the antibody, the label stuck to the filter.

This technique is very new and has not as yet been extensively tested, but if it lives up to expectations, experimenters will no longer have to inject whole animals with antigen in order to generate antibodies. Instead, they will simply order a library of antibody-producing bacterial cells and probe it with an antigen of interest. Because the antibody genes are present in bacteria, they may be subject to mutagenesis. There is therefore an opportunity to select for antibodies that have higher or lower affinities for any given antigen. The technique also opens up the possibility of easily obtaining human antibodies, something that was previously difficult to do. (Humans resent being used as antibody factories, for obvious reasons.) If it does become relatively easy to obtain specific antibodies, there should be a large increase in their use in the diagnosis and treatment of disease.

Vaccines

In spite of the widespread use of vaccines, infectious diseases still are a major source of concern. Not only do they distress the human population but they kill and sicken pets, farm animals, and wildlife, causing a significant impact on our everyday lives. Some diseases have succumbed to vaccination, but others remain refractory. And new conditions seem to arise regularly, witness AIDS, that require the development of new vaccines. Recombinant DNA has already had an influence in this area, and it promises to have an even greater impact as time goes on.

Bolstering the body's own response to an invading microbe by immunization dates back to Edward Jenner's work in the eighteenth century. Over the years, three different immunization methods have generally been employed: the injection of live organisms, dead organisms, and subunits of disease-bearing agents. Each of these has certain disadvantages. Live organisms are often extremely effective in eliciting an immune response because they proliferate and persist. But there is always the danger that they may mutate to produce either the original virulent strain or to take on some other undesirable characteristic. Inactivated virulent organisms often aren't as effective in provoking immunity because they're frequently quickly cleared from the body. Moreover, they must be extensively tested to be sure that they are completely free of live material. The injection of subunits (an immunogenic part of the disease organism) lacks this last drawback, but parts of an organism often are less immunogenic than the whole. In addition, since the subunits must be purified, the cost of subunit vaccines may be higher than the other types – an important consideration if the vaccine is to be used in Third World countries.

Recombinant DNA overcomes many of these problems. First, since the essence of recombinant DNA techniques involves the expression of foreign genes in *E. coli*, it would appear easy to make a subunit vaccine by taking the appropriate gene from a disease organism, putting it into a plasmid, and expressing the protein in large amounts. This strategy was used to produce one of the first recombinant DNA success stories: a vaccine against hepatitis B.

But all is not perfect. There are two general difficulties.

First, as noted several times, heterologous expression systems sometimes do not provide proper post-translational processing of proteins. And it turns out that some of these post-translation steps may be very important for maximizing the immune response. For example, in the case of hepatitis, *E. coli* was found to be an unsuitable host for expression of hepatitis B surface antigen, a protein that carries attached carbohydrate residues and which is the major viral antigen. Yeast proved to be a more successful host, and the yeast-derived product has been approved for use with humans.

Second, the production of antigens in bacteria or any other host has the same limitations that any subunit immunogen has: The product often is less immunogenic than the whole disease organism. Where this occurs, another strategy has been employed.

Vaccinia virus

Vaccinia (a weakened poxvirus similar to the smallpox virus) is responsible for virtually eradicating smallpox from the world. It is a double-stranded DNA virus with a large genome (about 185 kilobase pairs and about 200 genes), which replicates in the cytoplasm of infected cells. Like the baculovirus described above, vaccinia hasn't been well characterized molecularly: Only a few genes have been sequenced, and few mutants are known.

In the early 1980s, it was suggested that a recombinant vaccinia virus harboring a passenger gene coding for an antigen from a disease organism would have many of the advantages of a live vaccine (chiefly persistence, because of the ability of the organism to replicate). Yet because vaccinia is benign, it might offer few of the disadvantages of injecting a live disease organism. In addition, the virus is stable to freeze-drying, and it can be administered by scratching it onto the skin.

Because of the lack of molecular sequence information, the same strategy used to introduce genes into the baculoviruses was employed for cloning into vaccinia. A plasmid was constructed containing a portion of the vaccinia genome surrounding a suitable cloning site. The gene for a foreign protein (a suitable antigen) was then inserted into this site, and the recombinant plasmid was introduced into tissue culture cells (via transformation) that had already been infected with vaccinia. During the course of virus replication, the foreign gene sometimes becomes integrated into the virus by homologous recombination with the viral sequences on the plasmid. Recombinant viruses containing a passenger gene can be identified by including a selectable gene on the plasmid in between the vaccinia recombination sequences. Unlike lambda bacteriophage, it doesn't seem to much matter how much passenger DNA is introduced.

Several antigens have already been inserted into vaccinia using this method, including rabies virus glycoprotein (the term glycoprotein indicates that the protein has attached carbohydrate residues) and influenza virus hemagglutinin. These have already proved effective in evoking antibodies and immunity. Currently, a number of groups are trying to determine whether an effective AIDS vaccine can be produced with this approach.

Agricultural applications

Transgenic animals

Animals that have one or more foreign genes inserted into their germ line are called **transgenic**. They can pass the gene on to their offspring, and it can be inherited as a Mendelian trait. For a long time, the only transgenic mammals were mice, the most well-known example being a large animal pictured on the cover of the magazine, *Nature*, that carried a rat growth hormone gene. This mouse and other transgenic mice must have whetted the appetite of the agricultural community. With images of made-to-measure cattle, sheep, goats, and poultry in their heads,agricultural scientists have moved forward in an effort to introduce similar genes into domestic animals.

Before embarking on a discussion of the potential impact of gene transfer, it may be worthwhile to review some of the basic methods for obtaining transgenic mammals. Transgenic mice (the model system for most of the work in this area) have been produced in three different ways.

   • The first means, the classic method, is to manually inject (with the aid of a microscope and micromanipulator) hundreds to thousands of copies of a gene into the pronucleus of a recently fertilized egg. The eggs are cultured overnight in an appropriate medium, and embryos are transferred to a properly prepared female (one that had been mated with a sterile male). Some of the resulting offspring, typically on the order of a few percent, will inherit the newly introduced gene and pass it on to succeeding generations.

   • A gene may also be placed in the germ line of mice by first getting certain tissue culture cells to take up foreign DNA. The trick to this technique is to use the right cell type. In most cases, embryonal stem cells are used. They are a line of early embryonic cells that can grow in culture. When mixed with normal cells of the growing embryo, they are capable of giving rise to all the cell types of an organism. Sometimes they differentiate into germ cells. If these contain integrated foreign DNA, they are capable of passing on an introduced gene to their offspring.

   • A third way of obtaining transgenic mice is by the use of retroviral vectors. Retroviruses (the HIV virus is a familiar example) are RNA viruses that replicate via a DNA intermediate. The viral DNA carrying a foreign gene integrates into the genome of its host where it functions to produce RNA, just like any other gene. In practice, transgenic animals have been obtained by infecting early mouse embryos with recombinant retrovirus vectors.

To date, all transgenic mammalian animals other than mice have been produced by injecting DNA into pronuclei. Transgenic rabbits, sheep, and pigs have been constructed, many carrying a growth hormone gene. There have been some problems. Pigs with high levels of growth hormone have not behaved as their mouse cousins: They do not show a rapid increase in weight. Sheep that make large quantities of recombinant sheep growth hormone gene tend to be unhealthy, often dying within a year of birth.

However, there have been some successes too. The most successful farm animal studies to date have utilized mammals as so-called bioreactors. By hooking human protein coding genes (a clotting factor, for example) to a sheep regulatory sequence that causes transcription in mammary tissue, Scottish scientists have been able to obtain transgenic sheep that make milk containing human proteins. Similarly, a group in Switzerland has succeeded in getting rabbits to produce human interleukin-2 in their milk under the control of a rabbit casein regulatory DNA sequence. In both rabbits and sheep, yields have so far been relatively low. This is a technology that promises much, but has yet a long way to go.

Plant transformants

Dicotyledonous plants (broad-leaf plants with two leaves in their sprouts) are subjected to a disease called **crown gall** which is characterized by the formation of large tumors at the junction of stem and shoot. The disease is caused by the bacterium *Agrobacterium tumefaciens*. In turn, the agent in *Agrobacterium* that is responsible for the disease is a large plasmid (200 to 250 kilobase pairs) called the **Ti plasmid** (for tumor inducing), a portion of which enters the cells of the plant (called **T DNA**, for transforming or transferred) and integrates into the plant genome. Once resident in plant chromosomes, the T DNA codes for enzymes that produce several unusual substances called **opines** and for proteins involved in the synthesis of plant growth hormones.

It is evident that *Agrobacterium* is a genetic engineer par excellence. It actively transfers a set of useful bacterial genes into the genetic apparatus of plants. The opine genes that are introduced end up making a set of enzymes that produces specific nutrients for the bacterium. At the same time, the growth hormone genes act to increase the mass of the affected tissue. Molecular analysis has shown that the bacteria have been cunning enough to have attached eukaryotic regulatory sequences to these genes so they can be properly used by the plant. People mindful of the cleverness of this system, of course, have subverted it to their own ends.

It turns out that while the T DNA is populated with at least a dozen genes, only the direct repeats at its borders (24 base pairs long) are important for its integration into the plant genome. Given that information, a plasmid was constructed in which the border sequences were left intact, and virtually all of the intervening genes were deleted and replaced with plasmid DNA. This construct, a **disarmed** Ti plasmid, does not produce crown gall tumors. It is, after all, missing the hormone and opine synthesizing genes. But upon infection, a piece of DNA is still transferred into the plant. In fact, it has been found that any gene that is placed between the border sequences is effectively integrated into the plant genome.

The next question that genetic engineers faced was how to make these disarmed plasmids into effective cloning vehicles. Unfortunately, even without the genes between the border sequences, the disarmed Ti plasmid is still far too big to function as a good cloning vehicle. There are too many restriction endonuclease sites in the remainder of the plasmid. To move genes into the Ti plasmid, like the baculoviruses and vaccinia viruses described above, use is made of recombination between it and a "donor vector" to produce a new vector capable of carrying a foreign gene. Alternatively, a binary Ti vector system can be used. The binary vector consists of two plasmids, each capable of replicating in *Agrobacterium tumefaciens*. One plasmid is large and carries the genes required to direct T DNA into the host. The other is small and carries the foreign gene, the one that is to be inserted into the plant genome, between the border sequences. In addition, the smaller plasmid may bear a marker so that transformed cells can be selected.

Once a vector has been constructed, a variety of options are available for carrying out the actual transformation. One widely used technique is leaf disc transformation. Here small discs are cut out of a young leaf with a paper punch, and the wounded tissue is cultivated with *A. tumefaciens* bacteria containing an appropriate cloning vector. After a suitable period, the bacteria are killed, and the discs transferred to paper filters on feeder plates for two days. Then the filters are put on "shooting medium" containing an antibiotic that selects against non-recombinant containing cells and encourages surviving cells to form shoots. After selection, the shoots are put on rooting medium, and after roots form, the small plants are placed in soil. This technique has worked with tobacco and tomato plants.

Transformation of plants is a fast moving field. Besides the use of Ti plasmids, a variety of plant viruses are being developed as cloning vectors. In addition, transformation has been achieved by shooting tiny tungsten pellets coated with DNA into plant cells. This too is an area where much progress is being made, and we do seem to be on the verge of a breakthrough.

12

The computer in molecular biology

Within the last few years, digital computers – particularly personal computers and workstations – have become as familiar and essential in the genetic engineering laboratory as spectrophotometers and centrifuges. Computers are useful in many ways. Suppose, for example, that a genetic engineer has just sequenced a fragment of DNA. How does she know that it hasn't already been published by someone else? Imagine the difficulty of manually looking through all the sequences reported in the literature for a succession of A's, T's, G's, and C's that match similar sequences that have been found in her laboratory. Computers, because they are very patient and exact and never seem to get tired or bored, do this job much better than people.

As an example, suppose a genetic engineer sequences a protein coding region of a previously uncharacterized gene. The computer can translate the triplet code from this DNA sequence into an amino acid sequence. Moreover, it can search the sequences of all known proteins to find out any that have areas similar to the one that has been newly obtained. This procedure is called a **homology search**, and it has become an important part of every sequencing project. In recent years, the molecular biology community has become quite adept at identifying areas of amino acid sequence within proteins that suggest particular functions. This becomes particularly important when a gene of unknown function is being characterized.

There are still other useful functions that computers serve in genetic engineering. They find restriction sites in sequences and do restriction mapping. They draw pictures of plasmid constructions, marking the restriction sites and various fragments in different colors and patterns if necessary. They keep track of sequencing and cloning projects.

Because of the importance of computers in the laboratory, it's important that the student of recombinant DNA know at least some of the basics of computer form and function. For this reason, some of the programs and databases that are widely available are described. These allow molecular biologists to carry out both mundane tasks as well as more sophisticated tasks.

Theoretical considerations

Computers, like molecules, come in two basic varieties. Personal computers (or personal workstations), the first type, are machines that are dedicated to a single user. The other kind of computer is a shared device – a machine that is designed to service many people at the same time. (The distinction between these two classes has blurred a bit with the advent of computer networks. So while personal computers are generally still designed with a single user in mind, that user can be hooked up to other single users, sometimes in large arrays.)

Personal computers

When the personal computer revolution arrived, there were many brands of personal computers in the marketplace. Over the years, there has been a rapid selection process, and today only a few kinds of popular personal computers are available. Of these, only two are widely found in the genetic engineering laboratory: IBM computers (and compatibles or clones) and Apple Computer's Macintosh. Both of these types can be used successfully in the biotechnology laboratory.

Multiuser computers

In contrast to the small size and intimacy of the personal computer, the popular picture of the multiuser computer is of an awesome machine kept guarded in a climate-controlled room, accessible only to a few of the priesthood. This concept is changing as multiuser machines become smaller and somewhat more accessible, but it is still true that many of these large computers must be maintained by professionals.

Multiuser computers, as the name implies, can be used by more than one person at once. Actually, the computer cannot really support multiple users simultaneously. It just appears that way because the computer is so fast that it can perform millions of operations in the time that it takes a user to make two successive keystrokes. While multiuser computers should not be intimidating, they do require some significant study before they may be used optimally. Plan on spending at least one week learning the command and file structures before attacking a major project. The same goes for personal computers, although with them the learning period may be somewhat shorter.

The computer in the laboratory

There are five general areas where computers are useful in the recombinant DNA laboratory:

- For analyzing protein and nucleic acid sequences

- For communicating with colleagues

- For treating data with statistical and mathematical tools

- For data acquisition and instrument control

- For instructional purposes, making illustrations and slides, and writing and publication

Many of these uses are self-evident. The first area is emphasized here.

Nucleic acid and protein sequence analysis

GenBank

As mentioned above, one of the more common chores that needs to be done after a new sequence is determined is to find out if there is another sequence to which it is similar. This task is facilitated by collections of DNA sequences that are available to the public. One of these is **GenBank,** a United States government sponsored repository and distributor of nucleic acid and protein sequences. Sequences are compiled and annotated (mostly from the published literature, but more recently from direct submissions) by a group at the Los Alamos National Laboratory. The data are then transmitted to IntelliGenetics, a firm in California that specializes in molecular biology computer programs. IntelliGenetics maintains and distributes the data in the form of databases, organized compilations of sequences, references, and annotations. In addition, GenBank includes several programs that allow users to locate the sequence of a specific gene in the database and another to search for entries that are similar to a specific sequence.

The number of DNA sequences in GenBank's repository is very large and growing very rapidly. For example, in 1983 and 1984, the rate of growth was approximately one million bases every nine months. Near the end of 1987, it took only nine weeks for the repository to grow by the same amount. As of March 1990 (Genbank release 63.0), there were some 41,143 sequences representing more than 40,127,752 bases worth of sequence in the bank. It is expected that the repository will grow even faster in the near future as large DNA sequencing projects begin to produce data.

The GenBank DNA listings are organized into subdivisions that include primates, rodents, other mammals, other vertebrates, invertebrates, plants, organelles, bacteria, structural RNA's, viral sequences, bacteriophage, and synthetic DNA's. There are additional subdivisions that allow users to search unannotated sequences and newly arrived sequences. GenBank can also be used to search the EMBL and protein databases that are mentioned below.

A typical entry in the GenBank format is shown below. It happens to be an invertebrate sequence, from the fruit fly *Drosophila melanogaster*.

```
LOCUS        DROADHC       2721 bp     DNA                pre-entry
05/01/84
DEFINITION   d.melanogaster alcohol dehydrogenase gene (allele adh-
s), consensus sequence.
ACCESSION    K00651
KEYWORDS     alcohol dehydrogenase; dehydrogenase; polymorphism.
SOURCE       drosophila melanogaster genomic dna; strains wa-s, fl-
1s, af-s, fr-s, fl-2x, and ja-s.
ORGANISM  Drosophila melanogaster
             Eukaryota; Metazoa; Arthropoda; Insecta; Diptera.
REFERENCE    1   (bases 1 to 2721)
   AUTHORS   Kreitman,M.
    TITLE    nucleotide polymorphism at the alcohol dehydrogenase
locus of
             drosophila melanogaster
  JOURNAL    Nature 304, 412-417 (1983)
COMMENT      this consensus sequence is based on six adh-s alleles.
FEATURES         from  to/span     description
```

```
        pept        841       939      alcohol dehydrogenase
                   1005      1409
                   1480      1746
BASE COUNT          833 a     630 c     587 g     671 t
ORIGIN
INVERTEBRATE:DROADHC   Length: 2721   2-DEC-1986 10:27   Check: 7444
..
1      GTCGACTGCA CTCGCCCCCA CGAGAGAACA GTATTTAAGG AGCTGCGAAG
51     GTCCAAGTCA CCGATTATTG TCTCAGTGCA GTTGTCAGTT GCAGTTCAGC
101    AGACGGGCTA ACGAGTACTT GCATCTCTTC AAATTTACTT AATTGATCAA
151    GTAAGTAGCA AAAGGGCACC CAATTAAAGG AAATTCTTGT TTAATTGAAT
```

The sequence goes on for all 2721 base pairs.

Most of the meanings in the entry are self-evident, but a few points may need clarification or emphasis. For example, the features section tells the reader (or a computer program accessing the data) important information about the DNA sequence. In the example above, the peptide-coding regions are indicated as beginning at nucleotide 841 and going to nucleotide 939. Then there is an intervening sequence, and the protein-coding starts again at position 1005 and extends to base 1409, and so on.

How is it possible to communicate with GenBank? There are several ways. Using a modem hooked up to a personal computer, anyone can call up GenBank and capture sequences from the data bank directly. Users are granted seven minutes of free access time. Alternatively, an individual on a shared system connected to the outside world can send an electronic mail message via one of the international computer networks (like BITNET or Internet). As of this writing, talking to GenBank via electronic mail allows an interested party to retrieve a given sequence from the database or to search the database for the presence of a sequence similar to one that is typed in. If neither of these modes satisfy, computer tapes (or disks) containing the entire collection of sequences can be purchased. The tapes are supplied for both personal computers and shared systems. Armed with all this information, genetic engineers can use commercial programs or ones that they write themselves to ask questions about the sequences that are in the collection.

EMBL

A second repository of sequence data is the European Molecular Biology
Laboratory (EMBL), the first internationally supported central nucleic acids
sequence resource. The EMBL data library was established in 1980 and, under
the leadership of Greg Hamm, the first sequence collection was released in 1982.
EMBL gathers sequences from the published literature (and from submissions
directly from authors) as does GenBank. The two repositories also exchange
data. The data in the EMBL data bank are organized differently than that in
GenBank, as shown below.

```
ID   DMADHS       standard; DNA; 906 BP.
XX
AC   V00199;
XX
DT   16-AUG-1982   (first entry)
DT   15-APR-1983   (minor modifications)
XX
DE   Drosophila gene for alcohol dehydrogenase
DE   (ADH; Alcohol:NAD+ oxidoreductase, [EC 1.1.1.1])
XX
KW   reductase; dehydrogenase; oxidoreductase.
XX
OS   Drosophila melanogaster (fruit fly, drosophila, Fruchtfliege)
OC   Eukaryota; Metazoa; Arthropoda; Insecta; Diptera.
XX
RN   [1]   (bases 1-906)
RA   Benyajati C., Place A.R., Powers D.A., Sofer W.;
RT   "Alcohol dehydrogenase gene of Drosophila melanogaster:
RT   Relationship of intervening sequences to functional
RT   domains in the protein";
RL   Proc. Natl. Acad. Sci. USA 78:2717-2721(1981).
XX
FH   Key          From     To       Description
FH
FT   CDS           1       99       reading frame 1st part
FT   CDS          165      569      reading frame second part
FT   CDS          640      903      reading frame third part
FT   IVS          100      164      first intron
FT   IVS          570      639      second intron
XX
SQ   Sequence     906 BP;  192 A;  260 C;  173 T;  210 G;  71 -.

     EMBL:DMADHS  Length: 906  2-DEC-1986 10:46  Check: 98  ..

   1  ATGTCGTTTA CTTTGACCAA CAAGAACGTG ATTTTCGTTG CCGGTCTGGG

  51  AGGCATTGGT CTGGACACCA GCAAGGAGCT GCTCAAGCGC GATCTGAAGG
```

The sequence goes on for 906 base pairs.

Notice the differences in format between the two databases. Also note that, while the gene is the same, the sequence comes from a different source and therefore starts at a different position.

PIR (Protein Identification Resource)

The National Biomedical Research Foundation (NBRF) in Washington, DC, has maintained a database of protein sequences for many years. The original purpose of collecting such sequences was to use them to study the evolutionary interrelatedness among proteins. Protein sequences from the NBRF and two other sequence collection centers are accessible through the Protein Identification Resource (PIR), a United States Government funded program. PIR consists of a computer, several databases, and associated software.

All three of these databases are widely distributed. But many organizations do not purchase the tapes (or disks) directly from EMBL, or GenBank, or the NBRF. Nor do most investigators access the databases directly using their computers. Instead, most of the genetic engineering community generally purchases a package of nucleic acid sequence analysis computer programs. Included in the price is a subscription for periodic updates of the databases. Many of these packages are available, and one is discussed below. It runs on a multiuser computer.

GCG

The GCG package of computer programs originated in the Department of Genetics at the University of Wisconsin in 1982. In late 1989, John Devereux, who was (and is) largely responsible for developing and supporting the programs, formed a company – called Genetics Computer Group, Inc. (GCG) – that now sells the software. Over the years, the package has been through five major revisions and is now used by more than 15,000 scientists worldwide. At this time, these programs run only on VAX computers (a popular multiuser minicomputer made by Digital Equipment Corporation).

The GCG package includes some 90 "software tools" – relatively short and fairly simple programs that do small specific tasks. The advantage of this approach (as opposed to large programs that attempt to do everything) is that it is fairly easy to learn to do any one task. This is an important consideration because most users put much effort into the initial learning process, solve the problem they were working on, and then forget how to use the program months later when the same or similar problem arises again. Since the GCG programs are easy to learn, the time spent in this process is much reduced compared to that of many other packages.

The tools of GCG are divided into 15 groups that include the following:

Sequence editing - edits nucleic acid or protein sequences. GCG also has programs that are useful for changing a sequence format from that of one database to another.

Sequence comparison - aligns and compares two or more sequences. One program in this group performs a graphical comparison of two sequences.

Restriction mapping - finds the location of restriction sites throughout any DNA molecule.

Nucleic acid secondary structure - finds self-hybridizing structures. This is useful for determining the three-dimensional structure of certain small RNA molecules.

Pattern recognition and composition - looks for protein coding regions in DNA, finds start and stop codons, and calculates the frequency of appearance of the various codons in a coding sequence.

Printing and publication - readies data for publication by making attractive plots and print-outs.

These are just some of the many capabilities of GCG. The program can be purchased from Genetics Computer Group, Inc., University Research Park, 575 Science Drive, Madison, Wisconsin 53711.

Gene Construction Kit

Many genetic engineering cloning experiments require the assembly of complex arrangements of multiple genes. Regulatory regions from one gene may be appended to another, and parts of genes may be deleted, inverted, or transposed.

The computer in molecular biology

A small company called Textco, Inc. (27 Gilson Roa[d], [Hanover New] Hampshire 03784) has developed a useful persona[l computer program,] **[Gene] Construction Kit,** that allows users to simulate [gene cloning and follow] the progress of their assembly.

Gene Construction Kit is a complex program with many capabilities. It requires some training to discover what it can do how it works. However, the graphic features of the program and the so-called user-friendly interface of the Macintosh, make it relatively easy to use. And once learned, it is not easily forgotten. A few of the program's functions are described below, using illustrations taken from the tutorial.

Gene Construction Kit can simulate very complicated cloning experiments. For example, the program can display on the screen a graphic depiction of one or more plasmids (or any other cloning vector, for that matter). With the aid of a mouse, a pointing device that comes with the Macintosh, any segment (for example, a sequence between two restriction sites) can be selected and then excised. Cut DNA fragments may be pasted into a second site (on the same or a different plasmid) in either direct or inverted orientation. In the example shown at the right, a *Hin*dIII fragment of the Alcohol dehydrogenase (Adh) gene shown on pages 130 and 132 has been inserted into the *Hin*dIII site of pBR322. As shown, once a piece has been pasted, the recipient plasmid grows larger by the required amount. Selected pieces can

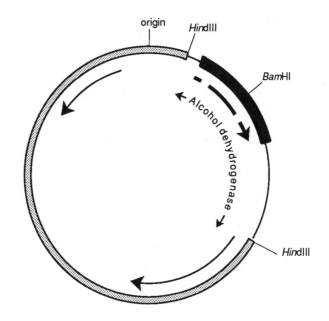

be marked by color or by pattern so that they may be followed in subsequent steps. In the example shown on the previous page, the coding portion of the Adh gene has been marked in black and is underlined by an arrow divided into three parts. Other parts of the vector (the tetracycline and ampicillin resistance genes) have been underlined by thin arrows.

Remarkably, the program not only keeps track of the ongoing construction project pictorially, it also allows the user to examine and even modify the DNA sequence as the construction proceeds. For example, suppose a fragment of DNA is excised from one plasmid, assigned a red color, and pasted into another. If the sequence view of the new construct is brought forward, a red DNA sequence can be observed inserted at the appropriate spot.

One other capability of the program merits mention. Most DNA constructions are checked by restriction mapping. For instance, the presence of an insertion can be detected by an increase in the total size of a plasmid or a particular restriction fragment. Gene Construction Kit has the very useful ability to **simulate** the restriction mapping of a construct. For example, if a piece of DNA can be inserted at a site in two different orientations, each orientation may often be distinguished by its restriction map. In the example shown at the right, the program has simulated the digestion (with *Bam*HI) and electrophoresis of the construct shown on the previous page. The left most lane shows the bands expected from a *Bam*HI digestion. The right most lane simulates the bands from a digestion of the same insert, but turned around 180°. This kind of analysis allows the genetic engineer to anticipate the results of restriction mapping and to determine which enzyme would best distinguish between the two orientations.

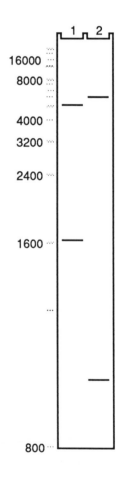

Summary

The computer is an indispensable tool in the genetic engineering laboratory. It can be used for a multitude of purposes, including finding restriction sites, determining homology among sequences, managing sequencing and cloning projects, and translating DNA sequences into protein sequences.

Repositories of nucleic acid and protein sequences have been formed to aid the genetic engineering community. GenBank, EMBL, and PIR are the major examples. They allow a genetic engineer to determine whether a newly found sequence is homologous to one already known.

13

Ethical considerations

This section is well outside my area of expertise: I can make no claim to being well versed in ethics. Nevertheless, I hold certain views on some of these matters, and although my views are admittedly biased because I've been a professional scientist for over 20 years, I am going to present them despite my lack of training in ethics. At the very least, I can claim to know something about the technical issues involved in the debate. I have heard others discuss these matters in the public forum without much of an understanding of the underlying science. While they are entitled to do so, it seems logical that in questions related to technical matters, an understanding of the basic science and technology is critical. In fact, that is one of the reasons that I have written this book.

The Frankenstein's monster scenario

What happens if you cross a gorilla and a parrot? The answer to this old joke is, "I don't know, but if it talks, you'd better listen." Of course, the real life answer is that nothing happens. Gorillas and parrots don't mate, and if somehow you were to transfer the sperm of one to the egg of another, nothing much would result. That's because mammals and birds are, obviously, very different from one another. Their genes have been subjected to millions of years of largely separate natural selection. In effect, each organism's set of genes has been designed to interact properly within that organism. If a set of one and a set of the other were to be mixed by fertilization, they would send the wrong signals to each other, the hybrid wouldn't develop very far, and it wouldn't work if it did.

That's also why the great majority of mutations have detrimental, if not disastrous, results. Organisms have been fine-tuned by evolution so that most of their gene products fit like jigsaw puzzle pieces into the physiological and developmental process. Enzymes, for example, have been selected to work best with certain concentrations of substrates and under highly specific conditions of temperature, acidity, and salt concentration. Most mutations in genes that code for enzymes change these properties – almost always for the worst. Similarly, the signalling and response of genes – their regulation – is carefully fine-tuned. Genes must make the right products in the correct amounts and in the proper places, and if they don't, all sorts of untoward consequences result.

So what happens if a gene from some foreign organism is inserted into another? Is it likely to produce a monster? As expected from what has already been said, recombinant organisms are almost invariably weakened when compared to their nonrecombinant siblings. Rather than creating a mighty mouse or a super fly, in all the cases done so far the genetically engineered organism seems relatively unaffected or is worse off than it was before. A specific example of this has already been covered: the attempts to overexpress a growth hormone gene in sheep and pigs (see Chapter 11).

But what about the chestnut blight or Dutch elm disease or the European rabbit in Australia or the Mediterranean fruit fly in California? Aren't they organisms that have run amuck? What's to prevent transgenic organisms from doing the same thing?

Chestnut blight and the other examples cited are **exotics**, creatures that were taken from one location (where they apparently lived in reasonable balance with disease and predators) and brought to another. Released from their usual biological bonds – from their predators and parasites – and presented with unoccupied habitats and plentiful food, they multiplied almost unchecked. Most organisms used for genetic engineering couldn't be exotics. They're found all over the world already. Most are domestic or laboratory organisms, many of whom have a hard time surviving without help from humans. That isn't to say that domestic animals released into the wild haven't done harm in some instances. Feral goats and pigs are examples of such organisms that have had profound effects on certain ecosystems. But it might be argued that introducing a foreign gene into them makes them less likely to be threats to the environment. Or, to put it another way, the environmental danger – if there is any – appears to be not in the recombinant DNA that has been introduced into the organism but in the organisms themselves.

Does that mean that there is no danger of something going wrong – of something unexpected happening – in a genetic engineering experiment? Readers who are expecting a resounding "yes" at this juncture are going to be disappointed. I know of no self-respecting scientist who will give absolute guarantees about these matters. Human beings are not all-knowing. Every act that we do (and for that matter, everything we fail to do that we could do), whether as individuals or as a society, may have consequences that we cannot anticipate. Similarly, when a stray cosmic ray hits the genetic material of some organism and causes a mutation, something unexpected (and perhaps terribly detrimental) will occur. And when a germ cell of a particular organism forms and the process of recombination (a normal part of every generation in most diploid organisms) occurs, certain combinations of genes could be produced that may have terrible consequences. And when organisms mate and the nuclei from two germ cells comes together to form a new individual, the combination may have properties that will bring disaster.

What I have tried to argue in this section is that the process of introducing recombinant DNA into an organism is no more likely to produce accidentally a highly destructive organism than the natural processes that are going on around us all the time. And, at least in the immediate future, there are no new techniques on the horizon to change that likelihood. Further, as shown in Chapter 11, there are substantial and immediate benefits of genetic engineering. It seems to me that the trade-off between the risks and rewards of genetic engineering is a reasonable one.

Curing diseased people by fixing their genes

Putting genes into plants and domestic and laboratory animals is one thing. What about using genetic engineering to modify humans, in particular to cure human disease? There are two quite different procedures that are being debated under this heading, and they need to be discussed separately. The first procedure is to insert foreign genetic material into the germ line – into sperm or eggs – where they could be passed on to succeeding generations. The second procedure is to use genetic engineering to insert foreign genes into somatic cells – into any cells except sperm or eggs and their progenitors – in an effort to correct some genetic defect.

This last procedure – somatic genetic engineering – is not a serious ethical problem to my mind. Or, put another way, the ethical considerations of somatic gene transfer are in principle the same as any other medical technique. The essential questions are these: Does the process work? Will it yield a better quality of life for the treated individual? Have the patients (or their parents or guardians) given informed consent?

On the other hand, treating human genetic disease by purposely changing the genes that are passed on to succeeding generations seems to me potentially dangerous and ethically unsound. It's dangerous precisely because of the problems brought up at the beginning of this chapter: There is always the possibility of some unknown interaction between the introduced gene and the remainder of the genome. And the consequences of this ignorance fall onto succeeding generations – on innocent children or fetuses. There is no possibility of informed consent. On the other hand, if the genetic engineer and the physician make a mistake in somatic therapy, the person who suffers is the patient, who should have been appraised of the dangers before the therapy was begun.

Everything you wanted to know about anybody –The new diagnostics

Suppose you a child has a genetic disease – one that could be cured by diet or a similarly straightforward therapy. Wouldn't the parents want the disease diagnosed? Or to take another example, perhaps two people are getting married. Wouldn't it be beneficial to know what chance their offspring have of inheriting one or several harmful genetic traits? Armed with the procedures of PCR, Southern blotting, RFLP mapping, and DNA sequencing, laboratories have already developed the ability to provide a very detailed look at our genetic material. Should these techniques be applied to all of us? What harm might it do to have a complete picture of our genes? After all, the procedures don't insert new genetic material into cells. In fact, no changes of any kind are made. We are only being provided with information about what is already there. Where is the controversy?

Unlike the first two sections of this chapter where I've tried to take a strong position for the use of genetic engineering in most situations (except for the introduction of genes into the germ line of humans), I'm much less sanguine about the prospect of widespread use of DNA diagnosis. I can foresee many dangers inherent in the technology, and I believe that we should be very careful about what we are getting into.

The root of the danger is that the information that is obtained by the new DNA diagnostics can be used to unfairly discriminate among people. For example, suppose your daughter is interviewing for a job. Her prospective employer asks for her DNA sequence (or a detailed RFLP map), which she carries around in her wallet on a small magnetic card. "I'm sorry," the interviewer says after appropriate computer analysis of the data. "We don't employ people with your combination of genes because it's been shown that they have a greatly increased risk of Alzheimer's disease, and we can't afford the insurance." Obviously, insurance companies can use these same arguments if they are armed with the same evidence. And so can schools and even governments.

Because there is a genetic component of behavior, other prospects of the new diagnostics are equally disturbing. Schizophrenia, some forms of manic depression, and some other mental illnesses are thought, at least in part, to have a genetic basis. Some people think that intelligence (or at least the ability to score points on an IQ test) is partially genetically determined. A genetic disposition toward some trait means that there will be a correlation between the presence or absence of some gene or genes with that trait. Perhaps it will be discovered, given enough data, that there are correlations between success (however defined) and some genetic polymorphisms, or between political leanings and some genes, or between loyalty and patriotism and some base changes in DNA. All these correlates may be of academic interest to some sociologist or psychologist doing a research project somewhere, but they also could be put to prejudicial use by some company or school or government.

I don't mean to suggest that knowing someone's DNA sequence is inherently bad. There certainly are circumstances, mostly medical, where such information would be of vital importance. The aforementioned case of prenatal (or perinatal) diagnosis of a birth defect or an enzyme deficiency is one situation where DNA diagnosis could be of immediate beneficial use. What must be emphasized is that the information to be garnered by the new technology should be private and used by choice by the individual whose DNA has been read. A person's DNA sequence, to my mind, should be protected by law from distribution to other parties without the individual's specific permission. Similarly, physicians who have obtained such information should treat it as they do any privileged material.

Summary

The introduction of foreign genes into organisms doesn't seem to be dangerous in its own right. The resulting organisms don't appear to be better able to reproduce than any others.

Attempts to remedy human disease by somatic gene therapy also don't appear to pose ethical dilemmas. But the purposeful introduction of genes into the human germ line does.

The diagnostic procedures that genetic engineering allows are very powerful. Before long, it will be relatively easy to correlate the presence of various DNA polymorphisms with disease and other hereditary characteristics. A person's genetic information should be privileged.

Glossary

Acrylamide (polyacrylamide) gels

Gels polymerized from acrylamide. Used in electrophoresis to analyze relatively small molecules of DNA and when high resolution is required.

Agarose gels

Gels prepared by dissolving agarose (a derivative of agar) in buffer. Used in electrophoresis of relatively large pieces of DNA.

Alkaline phosphatase

From *E. coli*, calf intestines, or bacteria. Removes 5' phosphate from DNA or RNA. Used during recombinant DNA constructions to prevent circularization of vectors without inserts.

Autoradiography

The detection of radioactivity by the darkening of an overlaid photographic emulsion. After Southern blotting, for example, electrophoretically separated DNA fragments are transferred to a filter and hybridized with radioactive probes. The filter is then placed on a sheet of X-ray film, and the individual bands are visualized by the dark bands that they form on the emulsion.

Bacteriophage

A virus that infects and often kills bacterial cells. See lambda phage and M13.

Blunt (flush) ends

The termini of double-stranded DNA molecules that are completely base-paired. Blunt ends contrast with sticky ends, which have unpaired single-stranded extensions.

cDNA (complementary DNA)

A DNA molecule that is complementary to an RNA molecule. Usually synthesized by the enzyme reverse transcriptase.

Clone

A set of genetically identical cells or organisms derived from a single individual by asexual processes.

Cohesive (sticky) ends

Single-stranded termini of double-stranded DNA molecules that are complementary to one another. Some restriction endonucleases cut DNA so that cohesive ends form, others form blunt ends.

Colony hybridization

A technique used to detect the presence of a cloned DNA segment in a bacterial colony.

Concatamer

Two or more identical sequences joined tandemly head to tail.

Cosmid

A plasmid that carries one or more *cos* sites from bacteriophage lambda.

DNA ligase (T4)

Repairs breaks in DNA, tying together 5' phosphates with a 3' hydroxyl group. T4 DNA ligase will splice blunt (flush) ends together.

DNA polymerase I (*E. coli*)

1. Polymerizes DNA (needs primer and template) in a 5' to 3' direction.
2. Is a 5' to 3' exonuclease.
3. Is a 3' to 5' exonuclease.

Used for labelling via nick translation.

DNase I

An enzyme that breaks double- or single-stranded DNA. In the presence of Mg^{++}, DNase I makes single-strand cuts in double-stranded DNA; in the presence of Mn^{++}, double-strand cuts are made. Used in characterizing nucleosomes and for making single-strand breaks in DNA prior to nick translation.

Eukaryote

Means "true nucleus". An organism that has its genetic material bounded by a membrane. Also characterized by the presence of membrane-bound organelles such as mitochondria or chloroplasts.

Exons

See introns.

Genome

The entire complement of chromosomal DNA of a given organism. Genomic clones are derived directly from the DNA, not from a cDNA copy.

Hybridization

Association by base pairing that occurs between two complementary strands of either DNA or RNA (or between DNA and RNA).

Introns (intervening sequences)

Segments of DNA that are transcribed but removed from the mature mRNA. The remaining pieces of RNA (exons) are joined together by splicing.

In vitro mutagenesis

Modifications made to isolated and purified genes by chemical or physical methods. Modifications include deletions, insertions, inversions, and transpositions.

Lambda phage

A temperate phage containing about 50kb of DNA. It is a useful vector for cloning large segments of DNA (up to about 20kb).

Library

A collection of cloned random fragments of DNA. Sometimes called a *gene bank*. Both cDNA libraries (a collection of the DNA versions of a group of RNA's) and genomic libraries are commonly encountered.

Linker

Self complementary oligonucleotides of defined sequence containing the cleavage site of one or more restriction endonucleases.

M13

A filamentous bacteriophage that carries a single strand of DNA.

Methylases

A number of enzymes that introduce methyl groups into DNA. These enzymes may be used to modify specific sites in DNA and prevent their being cut with certain restriction enzymes.

mRNA

RNA that serves as a template for protein synthesis. Called "messenger" RNA. Its sequence may differ from that of the gene from which it was transcribed because of the removal of introns and other processing events.

Mung-bean nuclease

An enzyme that works like S1 nuclease, but gentler in its activity.

Nick translation

Nick translation makes use of the 5' to 3' exonuclease activity of DNA polymerase I to move ("translate") a nick in DNA from one position to another. The polymerase begins synthesis of a new strand at a nick, degrading the DNA ahead of it. Nick translation is used to label DNA so that it can be used as a probe.

Northern blots

A technique for analyzing RNA. The RNA is run out on a gel and then transferred to a suitable substrate such as a filter. The RNA of interest is then detected by hybridization with an appropriately labelled probe. See Southern blotting.

Plasmid

An autonomously replicating, circular, extrachromosomal genetic element.

Polyadenylation

The addition of poly A to the 3' end of an RNA molecule by a mechanism that does not involve transcription.

Polynucleotide kinase

Transfers the γ-phosphate of ATP to the 5' hydroxyl of DNA or RNA. Used for labelling DNA and RNA (with 32P-phosphate labelled ATP) and for restoring the phosphate removed by alkaline phosphatase.

Primer

In DNA synthesis, a single-stranded DNA, often an oligomer, that serves as a starting point for polymerization of a second chain. The primer base pairs with the template (the other strand) and is extended by a DNA polymerase to form a complementary strand.

Probe

A labelled nucleic acid (most commonly with one or more radioactive atoms) that is used to detect a complementary nucleic acid by hybridization.

Promoter

In prokaryotes, a DNA sequence that is recognized by DNA-dependent RNA polymerase and the site of binding of that enzyme. The promoter regulates the rate of initiation of transcription. In eukaryotes, promoter sequences are regulatory genetic elements that do not directly bind RNA polymerase. Instead, they bind a group of proteins that is recognized by the polymerase.

Restriction endonucleases (Type II)

Enzymes that recognize a particular sequence (usually 4-8 nucleotides) in double-stranded DNA and cut in or near that sequence, leaving 5' phosphate and 3' hydroxyl ends. Used to cut DNA molecules into pieces of defined size with specific kinds of ends.

RNA-dependent DNA polymerase (reverse transcriptase)

Synthesizes DNA 5' to 3' using a primer and single-stranded RNA as a template

(DNA may also be used). Used in the synthesis of cDNA, first strand.

RNase T1

Cleaves RNA. Used to remove large pieces of RNA from DNA preparations.

S1 nuclease (from *Aspergillus*)

Single-strand specific endonuclease. DNA/DNA, DNA/RNA, and RNA/RNA hybrids are resistant. Used in "S1 mapping," RNA protection experiments, and cDNA cloning to remove noncomplementary areas of DNA/DNA, DNA/RNA, and RNA/RNA hybrids.

Shuttle vector

A plasmid that is capable of growing in two hosts.

Southern blotting

A technique originated by Ed Southern. Pieces of DNA are transferred from gels after electrophoresis to filters where DNA fragments may be detected using radioactively labelled nucleic acid probes.

Terminal transferase

Adds deoxynucleotides to the 3' OH end of DNA molecules. Used to add homopolymer tails onto DNA before ligation and for labelling.

Transcription

Synthesis of RNA directed by a DNA template using the enzyme RNA-polymerase.

Transformation

Uptake of purified DNA of any kind by cells.

Translation

Synthesis of protein directed by mRNA.

BIBLIOGRAPHY

General background

"Genes IV," by Benjamin Lewin. John Wiley & Sons, New York (1990).

"Molecular Biology of the Gene," by J.D. Watson, N.H. Hopkins, J.W. Roberts, J.A. Steitz, and A.M. Weiner. Benjamin/Cummings, Menlo Park, California (1987).

"Recombinant DNA: A Short Course," by J.D. Watson, J. Tooze, and D.T. Kurtz. Scientific American Books, W.H. Freeman, New York (1983).

Theory

"From Genes to Clones: Introduction to Gene Technology," by Ernst-L. Winnacker. VCH Verlagsgesellschaft mbH, D-6940 Weinheim, Germany (1987).

"Principles of Gene Manipulation – An Introduction to Genetic Engineering," by R.W. Old and S.B. Primrose. In: Studies in Microbiology, Third Edition, edited by N.G. Carr, J.L. Ingraham, and S.C. Rittenberg. University of California Press, Berkeley, California (1985).

"DNA Cloning," Volumes I and II, by D.M. Glover. IRL Press, Washington, DC (1985).

Lab manuals

"Molecular Cloning – A Laboratory Manual," Second Edition, by J. Sambrook, E.F. Fritsch, and T. Maniatis. Cold Spring Harbor Laboratory, Box 100, Cold Spring Harbor, NY 11724 (1990). *This is the standard against which all others are judged.*

"Current Protocols in Molecular Biology," edited by F.M. Ausubel, R. Brent, R.E. Kingston, D.D. Moore, J.G. Seidman, J.A. Smith, and K. Struhl. John Wiley & Sons, New York (1987).

"A Practical Guide to Molecular Cloning," by Bernard Perbal. John Wiley & Sons, New York (1988).

"Recombinant DNA Methodology," edited by. J.R. Dillon, A. Nasim, and E.R. Nestmann. John Wiley & Sons, New York (1985).

"Practical Methods in Molecular Biology," by R.F. Schleif and P.C. Wensink. Springer-Verlag, New York (1981).

"Advanced Bacterial Genetics – A Manual for Genetic Engineering," by R.W. Davis, D. Botstein, and J.R. Roth. Cold Spring Harbor Laboratory, Box 100, Cold Spring Harbor, NY 11724 (1980).

"Recombinant DNA Techniques: An Introduction," by R.L. Rodriguez and Robert C. Tait. Addison-Wesley, Reading, Massachusetts (1983).

"Nucleic Acid Hybridization – A Practical Approach," by B.D. Hames and S.J. Higgins. IRL Press, Washington, DC (1985).

"Gel Electrophoresis of Nucleic Acids - A Practical Approach," by D. Rickwood and B.D. Hames, IRL Press, Washington, DC (1982).

Computer tutorials

For several reasons, educational programs, especially those designed to promote understanding of college-level material, are not plentiful. First, effective and easy-to-use programs have been extremely difficult to write. That's partially because large programs in general are hard to write, even by people who program for a living. Educators, who might be expected to be most interested in developing educational materials, often don't have the time or inclination to become progressional programmers. Second, and more important, most software developers don't understand– perhaps no one understands – the unique advantages that computers can contribute to the learning process. There simply hasn't been enough educational research to pinpoint the role of the computer in the teaching process. Finally, textbook publishers – the people who would seem to be most suited to distributing and promoting educational software– tend to be conservative. Computers are a new medium. Questions arise concerning distribution,technical support, illegal copying, and reliability of the hardware and software that publishers find difficult to answer.

The first problem, the difficulty of writing software, is being addressed by a series of new languages and environments that are designed for use by the nonprofessional programmer. Claris Corporation's HyperCard, Silicon Beach Software's SuperCard and MacroMind's Director, for example, are ideally suited for easing some of the pain of software development. They seem particularly good for developing educational material. Already, these programs, and others like them have already brought forth a wealth of material. However, the problems of what to write, and how to make use of the unique properties of the computer, remain.

I've been intrigued by this new medium and have written two computer-based tutorials to help understand molecular cloning and genetic engineering. Both are intended for use with the Macintosh family of computers. One was developed with HyperCard, the other with SuperCard. They are extremely similar, their main difference being the platform on which they run.

While intended to complement this text, the two programs include some material that is not covered in the book and omit some sections that are. In addition, I've tried to write a program that is nonthreatening and informal. I've also attempted to make use of the unique properties of computers, their ability to animate processes and to interact with the user, rather than simply repeat the book in a different medium.

The HyperCard version of "Cloning Tutorial" can run on all Macintosh computers that are capable of running HyperCard. The program is in black and white and requires HyperCard. It's price is $30.

The SuperCard version of "Cloning Tutorial" can run on the Macintosh II family of computers. It requires 2 megabytes of memory and a hard disk drive. The program is in color, and will not run properly with a black and white or grey scale monitor. It's price is $60.

Both programs are available from SofeWare Associates, 17 Norton St., Edison, NJ 08820. Make checks payable to SofeWare Associates. Please allow two weeks for delivery.

Index